THE DYNAMIC BODY TISSUES

Cui dono lepidum novum libellum
Arido modo pumice expolitum?

Catullus, 87–54 BC

TO MY FAMILY

THE DYNAMIC BODY TISSUES

*An account of the chalone mechanisms and other influences
that control the epidermis and the various body tissues, and
of the failure of these mechanisms in cancer*

William S. Bullough
University of London

1983 **MTP PRESS LIMITED**
a member of the KLUWER ACADEMIC PUBLISHERS GROUP
BOSTON / THE HAGUE / DORDRECHT / LANCASTER

Published in the UK and Europe by
MTP Press Limited
Falcon House
Lancaster, England

British Library Cataloguing in Publication Data

Bullough, William S.
  The dynamic body tissues.
  1. Cancer cells
  I. Title
  616.99'407   RC207
ISBN-13: 978-94-011-6263-0   e-ISBN-13: 978-94-011-6261-6
DOI: 10.1007/978-94-011-6261-6

Phototypesetting by Georgia Origination, Liverpool

Contents

Preface   vii

Introduction
  1 The problems   3

The dynamic epidermis
  2 The epidermal messenger molecules   17
  3 Epidermal cellular homeostasis   40
  4 Epidermal cellular organization   54
  5 The dominance of the dermis   74

Homeostasis in other tissues
  6 The epithelial tissues   83
  7 Muscles and nerves   100
  8 The connective tissues   106

The failure of homeostasis
  9 Chalones and cancer   127

Envoi
  10 General summary   153

      References   167
      Index   179

Preface

The study of life is the study of tier upon tier of interlocking homeostatic mechanisms, and the main theme of this book concerns that part of the system that ensures cellular and tissue homeostasis and thus maintains tissue mass, tissue structure and tissue function in the adult mammal. The functional existence of any adult tissue depends on the continuing control of the states of differentiation of its cells. Just as in an embryo, where cellular differentiation is initiated and controlled by the genetic responses of the cells to specific messenger molecules, so also throughout the whole of adult life does the genetic activity of the tissue cells continue to be regulated by similarly specific messenger molecules. The process of differentiation does not cease in the embryo but remains as the essential factor which ensures not only the function but also the continuing existence of the adult tissues.

Thus the first problem considered here, the methods of control of cell production, cell function and cell death, is discussed in terms of the nature and mode of action of those messenger molecules which, from moment to moment, determine the states of differentiation of the tissue cells. Some of these messenger molecules are synthesized in adjacent tissues while others are synthesized within the responding tissues themselves. These latter substances are the chalones, each of which is specific in its action to the tissue that produces it, and each of which is of fundamental importance for the maintenance and proper functioning of that tissue. Through the control of its chalone mechanism a tissue is a self-promoting dynamic structure.

This leads naturally to an examination of the inadequate state of the chalone system in tumour cells, and so to a consideration of the potential value of chalones in the control of cancer.

A second problem also considered here concerns the manner in which the tissue cells arrange themselves in tissue-specific patterns. At each stage of its limited life each cell occupies a particular position within the tissue, which implies that as it ages a cell must move from place to place along a specific route. Relatively little is yet known of the ways in which such movements are controlled.

During the many years in which the foundations of this book have been laid many personal debts have been incurred, both in relation to the ideas that have been developed, and to the facilities and finances that have been needed. In particular I am deeply grateful for the stimulus of innumerable discussions and arguments with Dr. T. Rytömaa and Dr. E. B. Laurence, with the late Dr. W. J. Tindall and Dr. C. L. Hewett, and more recently with Dr. E. N. Mitrani and Ms. Johanna Stolze. In practical matters I am indebted for the support given by Birkbeck College (University of London), by NV Organon in Holland, and by Weddell Pharmaceuticals Ltd. in London. Also in the production of this book Johanna Stolze's help has been critical.

Introduction

If the Lord Almighty had consulted me before embarking upon Creation, I should have recommended something simpler.

Alfonso X, 1221-1284
King of Castile and Leon

Chapter 1

The problems

1.1	The problem of cellular differentiation	4
	1.1.1 Gene control in bacteria	4
	1.1.2 Gene control in mammals	5
	1.1.3 Gene control in adult tissues	6
1.2	The problem of cellular homeostasis	7
	1.2.1 Theories of mitotic control	8
	1.2.2 The critical experiments	9
	1.2.3 The chalones	9
1.3	The problem of cellular organization	10
	1.3.1 The epidermis	11
	1.3.2 The positions of cells	11
1.4	The problem of the breakdown of homeostasis	11
	1.4.1 The chalone mechanism and cancer	12
1.5	Misconceptions	12

The cells of every living organism are in a continuously dynamic but stable state of flux, both within and between the chemical, genetic and cellular levels. There is always a constant flow of molecules in and out of the metabolic pathways, controlled partly by chemical and partly by genetic feedback mechanisms, and in multicellular organisms there is also usually a constant flow of tissue cells which pass from their birth in mitosis to their death in cellular old age. It is the latter phenomenon of cell flow that is the main theme of this book, and because mammalian epidermis is easily accessible, and its cells are ideally stratified, with mitosis basally and cell death distally, this is the main tissue considered here.

However, it is already clear that the rules that have been found to govern cell flow in the epidermis apply equally to all other epithelial tissues and, with modifications, to tissues in general. The conclusions reached regarding the dynamic structure of adult epidermis also shed light, on the one hand, on the processes of differentiation in the embryo, and on the other hand, on the nature of the breakdown of control which leads to cellular chaos, or cancer, and which is more typical of the older organism.

1.1 The problem of cellular differentiation

In the adult epidermis, as in most other adult tissues, the cells have only two main activities open to them: they may indulge in mitotic activity and so add to the stock of cells, or they may mature, synthesize their tissue-specific cell product, and die. The cells either differentiate for mitosis, that is to say those genes are activated that specify the enzymes needed to drive the mitotic process, or they differentiate for tissue function, that is to say those genes are activated that in epidermis specify the enzymes needed for keratin synthesis. These two states of cellular differentiation are always mutually exclusive; the mitosis genes and the tissue genes are not activated simultaneously.

1.1.1 *Gene control in bacteria*

The concept of cellular differentiation was originally developed to describe the embryological origins of tissues and organs, but the most revealing information on the nature of the process has come from studies of bacteria.

In the prokaryotic cell all the genetic information is always readily available, and from moment to moment the pattern of genes that is activated is determined by the presence or absence of specific messenger molecules, called effectors, most of which originate in the cell's environment. These effectors are commonly metabolites as, for example, when a particular nutrient acts to trigger the activity of those genes that specify the enzymes needed for the digestion of that nutrient. When all the food has been digested the absence of any more effector molecules results in the genes again lapsing into inactivity.

Thus at any one moment bacteria of the same genetic make-up may be synthesizing widely different groups of enzymes, and this has led to the basic definition of differentiation, 'that two cells are differentiated with respect to each other if, while they contain the same genome, the pattern of proteins which they synthesize is different' (Jacob and Monod, 1963).

It is probable that this method of gene control, by the triggering actions of messenger molecules, will be found to be common to all the various groups of living organisms, except that in the metazoans the effector substances are likely to be not simple metabolites but molecules specialized for the purpose (Bullough, 1967). Another major difference is that while in bacteria all the genes remain available for activation as circumstances demand, in the differentiated mammalian tissues considered here the irrelevant genes are so firmly inactivated that no technique has yet been devised for their reactivation. In other words the bacterial genome is open and labile while the mammalian tissue genome is closed and stable.

1.1.2 *Gene control in mammals*

The mammalian fertilized egg contains a genome of which any part can be activated, as in a bacterium. During embryonic development the activation of those gene operons that initiate the syntheses of the tissue-specific enzymes on which tissue function depends is believed to be triggered by substances, commonly called inducing agents, which can be likened to bacterial effector substances. The difference is that the activation of one set of tissue genes is accompanied by the permanent closure of all the other sets of genes that are relevant to other tissue types. Thus a tissue once formed remains stable.

However, in an adult mammal a few cell types with some limited embryonic potential do remain. These are certain groups of stem cells in which the final steps in differentiation have not yet been taken, and which can therefore give rise to more than one kind of end-tissue. The best known are the bone marrow stem cells which, given the appropriate stimulus, undergo final differentiation, for instance, into granulocytes or erythrocytes (see 8.2.1), and the connective tissue stem cells which can differentiate, for instance, into chondrocytes or fibrocytes (see 8.1.3).

These end-tissues, like all other types of adult mammalian tissues, can never redifferentiate to give rise to any other type of tissue cell, even during tissue regeneration (Bullough, 1967; Caplan and Ordahl, 1978). There is, however, at least one possible exception to this rule; one type of epidermis is known that can undergo partial dedifferentiation. In any mammal the hair follicles are formed from fetal epidermis but thereafter, when the epidermis has become fully differentiated, the formation of new follicles becomes impossible. However, the epidermis that grows inwards to cover the bony pedicel, left when the antler of a deer is shed, thereby regains its ability to give rise to new hair follicles (see 5.1.3). Some influence coming from the pedicel induces a limited degree of epidermal de-differentiation (Goss, 1972), which lends support to a suggestion that, if only the right techniques can be discovered, the closed regions of the genome of mammalian tissue cells may be, at least partially, re-opened.

The closed and stable genome typical of adult mammalian tissues is evidently also typical of the tissues of most if not all the other higher groups of animals, such as the insects, but conditions in the higher plants are often different. Thus, as is well known, in many plant cuttings it is possible to induce cells that are already differentiated as part of the stem or leaf to respond to a changed environment by activating those parts of the genome that are needed for redifferentiation into root tissues. In such plants the closure of the irrelevant genes during differentiation is not so firm as it is in the higher animals.

1.1.3 *Gene control in adult tissues*

The active or potentially active genes in fully differentiated mammalian tissue cells can be classified into three main groups: those that specify the enzymes on which depend the essential metabolic pathways, such as glycolysis and respiration; those that direct the sequence of syntheses on which the mitotic cycle depends; and those that are necessary for tissue function. These three groups may be termed respectively the essential genes, the mitosis genes, and the tissue genes. The essential genes are evidently always active, although in cellular old age, when they are finally inactivated, the cell may continue to survive for some time through the activities of the remaining messenger RNA and the already formed enzymes. This final phase of cell life is particularly obvious in the distal epidermal cells with their dead nuclei and in the non-nucleated erythrocytes.

However, for present purposes it is the mitosis genes, which must be the same in all cell types, and the tissue genes, which are obviously peculiar to each type of tissue, that are of particular importance. The evidence is clear that these two sets of genes, or operons, do not function simultaneously; the activation of the one operon always involves the inactivation of the other. Thus any newly-formed tissue cell, such as a basal epidermal cell, still retains the ability to differentiate in either of two

6

different directions, and from what has been said it is logical to search for the effector molecules that initiate these two alternative genetic programmes. Two such types of substances are now known: the non-tissue-specific mesenchymal factor, which is synthesized by the connective tissue cells and which triggers mitosis in the adjacent epithelial tissue cells, and the tissue-specific chalones, which are synthesized in the tissue cells and which inhibit mitosis and promote maturation in these same cells.

There is also a third and more mysterious cellular activity, that of the ageing of the mature and functional tissue cells (Bullough, 1967, 1975). This ageing process is closely controlled, so that in each post-mitotic maturing cell a timing mechanism is set in motion, which accurately determines the life-span of that cell. All mature and functional tissue cells are programmed for death. Their life-span is tissue-specific, varying from 1 or 2 days in the intestine lining to 2 or 3 weeks in the epidermis and to several years in the liver (Bullough, 1972). Furthermore, in any one tissue a change in the rate of cell production, as for instance after tissue damage, is immediately matched by an equivalent change in the rate of cell ageing and hence of cell death. Thus in each tissue the rate of new cell gain is closely and directly linked to the rate of old cell loss.

There is some evidence that the rate of cell ageing may also be gene controlled, and that consequently it may be possible to speak of an ageing operon. Like the mitosis operon this would be common to all types of cells, only its rate of action being tissue-specific.

1.2 The problem of cellular homeostasis

Thus a typical mammalian tissue consists of cells which are involved in the mitotic cycle and cells which are mature and functional and therefore also ageing and dying. The precise balance that is always maintained between the rates of new cell gain and old cell loss ensures a stable tissue mass and an adequate tissue function. The various tissues show the widest range in mitotic activity, that is in the rate of cell flow through the tissue, from a maximum in hair bulbs and in duodenal mucosa to zero in the neurones and skeletal muscles. These last tissues, which are built by mitosis in the embryo and fetus, are incapable of any mitotic activity, their rate of cell flow to death is almost nil, and most of their functional cells survive as long as does the animal itself (see 7.1.2, 7.2.1).

The only significant variation on this basic pattern is seen in those few tissues which cannot maintain their own mass by mitotic activity, and which therefore must constantly recruit new cells by a terminal embryonic-type differentiation from a stem cell population. These are the connective tissues of the body; the stem cell population maintains its proper mass in the usual way (see 8.3).

The evidence is clear that in all tissues, including those that are mitotic only at an early stage in life and those that comprise the various connective tissues, there exists a homeostatic control mechanism which maintains the rate of cell gain in exact balance with the rate of cell loss. In cancer it is this mechanism that is distorted or broken.

1.2.1 *Theories of mitotic control*

The concept of the cybernetic control of physiological processes began more than a century ago with Claude Bernard, who stressed the importance of a constant '*milieu interieur*' and who developed the idea of the chemical messenger as the means of communication whereby this constancy is maintained. Shortly after 1900 there followed the era of hormone discovery, begun by Bayliss and Starling (1902), when it became fashionable to try to explain all physiological mysteries in hormonal terms. One outcome of this was the belief that mitotic control is achieved through the action of a 'wound hormone' (see Abercrombie, 1957). This 'hormone' was supposed to be synthesized in dying cells or damaged tissues to initiate the burst of mitosis by which the lost cells are replaced and the damage is repaired. The same mechanism was also supposed to operate in normal tissues, since it was believed that in them cells must be constantly dying as they became 'worn out', especially in those epithelia that are exposed to adverse conditions. This theory led to the belief that no cell will divide unless it is stimulated to do so.

However, when in 1946 the problem was re-examined, clear evidence was found for the existence of mitotic inhibitors, which 'prevent cell division from becoming excessive and unregulated', and for the fact that these inhibitors are tissue-specific in their actions, since 'each type of tissue is controlled independently of its neighbours' (Bullough, 1946).

Some years later when theoretical models of possible growth control mechanisms were beginning to be developed, the concept of mitotic inhibitors was again promoted. The first of these models were those of Osgood (1957), who primarily considered the blood cells, and of Weiss and Kavanau (1957), who considered tissues in general. They were agreed that the simplest theoretical control mechanism that would accord with the known facts would be a negative feedback mechanism based on the action of a tissue-specific mitotic inhibitor synthesized by the tissue cells themselves. According to Osgood a 'mature differentiating cell produces inhibitors of cell division', and the removal of such mature cells from the system 'leads to a more rapid rate of cell division by decreasing the concentration of the inhibitor'. Apart from calling this inhibitor an 'antitemplate', the model of Weiss and Kavanau was similar, as also were the later models of Iversen (1961) and Mercer (1962), who dealt specifically with the epidermis.

There thus arose a new belief that any cell will naturally continue to divide by mitosis unless it is prevented from so doing. On this view the theoretical mitotic inhibitors could be regarded as the most important chemical messengers in the body, since in their absence all tissue organization would collapse and growth would be explosive.

1.2.2 *The critical experiments*

In this situation it was necessary to devise experiments that would demonstrate clearly whether in an adult mammalian tissue the control of mitotic activity is achieved through a mitotic stimulant, such as a 'wound hormone', or through a mitotic inhibitor, as was theoretically postulated. This was finally achieved by Bullough and Laurence in 1960(a). Using the skin of mouse ears, which are so thin that epidermal damage on one side results in the passage of a mitotic stimulus to the undamaged epidermis on the other side, they confirmed that the control of mitosis operates in a tissue-specific manner, that a 'wound hormone' does not exist, and that the epidermis contains a mitotic inhibitor which acts only on epidermal cells (see also Iversen *et al*, 1974).

All the evidence that has subsequently been obtained has confirmed the conclusion that each body tissue synthesizes a mitotic inhibitory substance which acts to inhibit cell division only in that same type of tissue (Bullough, 1975; Iversen, 1981).

1.2.3 *The chalones*

This historical sequence of theories and experiments may give the impression that the concept of mitotic control by tissue-specific inhibitors was developed logically step by step (see also 9.4.3). In fact the various contributors worked very much in isolation, and it was only later, when all the scattered evidence had been pulled together, that the picture emerged as a whole and that the existence of a previously unsuspected system of chemical messengers became apparent.

In 1962 Bullough named these substances 'chalones', using a word coined in 1913 by Schäfer to distinguish an inhibitory chemical messenger from a stimulating chemical messenger, which Starling in 1905 had called a hormone. The word was derived from the Greek χαλαω, meaning to loosen or lower the sail in order to slow the speed of a ship, and since it was never accepted into general use, it was free to be adopted in its present more specialized sense.

A chalone was originally defined as a mitotic inhibitor 'produced by a tissue with the primary function of controlling the growth of that same tissue' (Bullough, 1962). This definition was later expanded as follows: a chalone is a chemical messenger that inhibits cell division, that is tissue-specific in its origin and its action, that is non-species-specific (the cod fish epidermal chalone is active on human epidermis), that is non-cytotoxic, and whose action is reversible (Rytömaa, 1976a).

However, although such a definition may be adequate for practical assay purposes, it seems probable that it may in time need some modification. In particular, since in all tissues the process of mitosis is so uniform, it is not obvious how the activation or inactivation of the mitosis genes can be controlled tissue-specifically.

To meet this problem one suggestion has been that a chalone may react with tissue-specific receptor sites on the outer surface of the cell membrane, and that this may trigger within the cell a non-tissue-specific secondary messenger system, which would then be the actual inhibitor of mitosis. There is indeed some evidence that chalone molecules are in some way associated with the outer cell membrane.

Another theory is that a chalone may not be primarily a mitotic inhibitor, but that its main function may be to activate those tissue-specific genes on which tissue function depends. Since tissue function and mitotic activity are mutually exclusive, the activation of the tissue genes may then lead by a feedback loop to the silencing of the mitosis genes.

1.3 The problem of cellular organization

No tissue exists as a haphazard mass of cells, some being in the mitotic cycle and others being functional. As new cells are formed they manoeuvre into particular positions, acquire particular shapes, and so produce the typical histological tissue pattern. Although the way in which cells move to form tissues and organs in embryos has often been described, the way in which newly-formed cells manoeuvre into position in adult tissues has been almost neglected until the recent spate of interest in the complex movements of certain post-mitotic epidermal cells.

1.3.1   *The epidermis*

The layered structure of mammalian epidermis has long been considered to be relatively simple, the newly-formed cells being simply pushed out of the basal layer to form a randomly-packed, distally-moving keratinizing cell mass. However, in 1969 MacKenzie demonstrated that in mouse epidermis the post-mitotic cells organized themselves into a subtle pattern of stacked columns, and since then the same pattern has been found in hamster and guinea pig epidermis, as well as in some types of human epidermis. In guinea pig epidermis a second pattern has also been found in which the post-mitotic cells overlap each other like the tiles of a roof (Bullough and Stolze, 1978). However, in most human epidermis and in all forms of epidermal hyperplasia no pattern is apparent.

1.3.2   *The positions of cells*

The broad generalization is now emerging that, in epithelial tissues, the mitotic-cycle cells tend to lie in contact with connective tissue, or with special regions of connective tissue, while the post-mitotic functional and ageing cells tend to lie away from such contact. In all types of epithelial tissue the cells, as they become post-mitotic and leave their basal positions, are evidently guided on their journey by factors that have not yet been adequately defined (see Abercrombie, 1980). In epidermis, where this distally-directed movement may be particularly precisely organized, there is also some evidence that each layer of distally moving cells may ensure an equal degree of keratinization and ageing by a system of cellular intercommunication. In non-epithelial tissues, such as the dermis, the situation is more complex (see 8.1).

1.4   **The problem of the breakdown of homeostasis**

In theory, a system of cellular homeostasis, which ensures that in a normal tissue the rate of new cell gain is always exactly in balance with the rate of old cell loss, could break down in either of two directions: if for any reason cell loss came to exceed cell gain the tissue would disappear, while conversely if cell gain came to exceed cell loss the tissue would continue to grow. The first possibility is seen, for instance, in the disappearing tail of a tadpole or of a human embryo; in an adult mammal it is also seen in the granulocyte and erythrocyte tissues, which in consequence must be continuously recruited from an undifferentiated stem cell population in the bone marrow. If ever it were to develop pathologically due to cell damage in small areas of tissue the result would be a 'negative tumour', which would simply vanish. For this reason, if for no other, it has never been reported.

The second possibility, an excess of cell production over cell loss, occurs naturally in a controlled manner in a growing animal, it occurs pathologically, but also in a controlled manner, in a healing wound or in a regenerating tissue or organ, and it occurs pathologically in an apparently uncontrolled manner in a growing tumour.

1.4.1   *The chalone mechanism and cancer*

A growing cancer is typified by a continuing excess of cell production over cell loss, by inadequate cellular maturation and tissue function, and by a disorganization of the cellular pattern typical of its tissue of origin. These changes in cellular behaviour derive from a particular pattern of cell damage that is permanent in that it is inherited by all the daughter cells, and it is obvious that this damage must at some point impinge on the chalone control mechanism of the affected cells. Indeed the belief that tumour cells are characterized by an inadequate chalone mechanism was implicit in the early theories of Osgood (1957) mentioned above, and it has been greatly strengthened in recent years by the results of a long series of experiments, particularly on epidermal carcinomata, various forms of myeloid leukaemia and a number of ascites tumours (see Chapter 9).

Certainly it can be confidently assumed that an understanding of the nature of the normal mechanism that determines cell numbers and tissue mass, and which is now believed to depend on the chalone mechanism, is an essential prerequisite for an understanding of the problem of cancer.

1.5   **Misconceptions**

The recognition of the existence of chalone mechanisms which control tissue-specifically the mitotic rate, and probably also tissue function and cell ageing, has developed steadily since about 1960. From time to time biochemists and cell physiologists have complained that no reality should be accorded to the chalones until they have been chemically characterized, which is a point of view that will not impress anyone who knows anything of the history of hormone research and usage. Biological evidence can be fully as impressive as biochemical evidence.

The cell physiologists have also confused the situation by insisting that the control of mitotic activity and of tissue function must involve a far more complex mechanism than is at first sight envisaged in the chalone theory. However, although the details of the control mechanism or mechanisms must indeed be complex, it does not follow that the trigger or triggers that activate the mechanisms need be anything but relatively simple.

The confusion has arisen for two main reasons. First, the cell physiologists have typically concentrated their studies on abnormal and aberrant cell lines, which have often been derived from tumours or have been 'transformed' by oncogenic viruses, and which have often been long separated from their tissues of origin when these are indeed known. Second, the problem of concocting the complex life-sustaining media needed by these cells has led to the cataloguing of a confusingly large number of so-called 'growth factors', which have come to be regarded as the normal controllers of mitotic activity. According to Holley (1975), these include 'fibroblast growth factor, epidermal growth factor, insulin, hydrocortisone, prostaglandin F, cyclic nucleotides, calcium ions, amino acids, phosphate ions, glucose and proteolytic enzymes' as well as 'peptides, antigens, lectins and lipopolysaccharides', and his conclusion is that 'the control of growth of mammalian cells will probably be found in the complex interactions of cells with these common types of materials'.

It is important to emphasize from the beginning that lists such as this are meaningless in terms of normal tissues in a normal animal. This particular list contains three main classes of substance: metabolites and hormones which are naturally present in the body in adequate and controlled concentrations; substances like proteolytic enzymes and antigens which are significant mainly or only in pathological situations; and substances like epidermal growth factor which so far have no known role *in vivo*.

Wessells (1977) has properly emphasized how false and misleading it may be to extrapolate to *in vivo* situations the results obtained from cell lines *in vitro*, and he has insisted that 'regulatory mechanisms operating in intact tissues and organs may be quite different from those working in a cell line in culture'.

A normal cell in a normal tissue in a normal animal is not short of any material for its survival, for its multiplication, or for its proper tissue function. All it needs is the instruction as to which set of genes should be activated, and it is now evident that instructions of this kind are most likely to be expressed in terms of the local concentration or effectiveness of one or more specialized messenger molecules such as those of the various chalones. It now seems probable that the many different chalones will prove to be a family of related substances.

The Dynamic Epidermis

Die wissenschaftliche Untersuchung der Natur strebt in den Einzelheiten das Allgemeine zu erkennen.

Karl Ernst von Baer, 1792-1876

Chapter 2

The epidermal messenger molecules

2.1	*Cell gain and cell loss*	19
	2.1.1 *The mitotic cycle*	19
	2.1.2 *The ageing pathway*	21
2.2	*The mesenchymal factor*	23
	2.2.1 *Mitotic stimulation*	23
	2.2.2 *The nature of the mesenchymal factor*	24
2.3	*The epidermal G1 chalone*	25
	2.3.1 *Mitotic inhibition*	25
	2.3.2 *The action of the glucocorticoid hormone*	27
	2.3.3 *The control of cell maturation*	28
	2.3.4 *Fetal and neonatal epidermis*	29
	2.3.5 *The nature of the G1 chalone*	30
2.4	*The epidermal G2 chalone*	31
	2.4.1 *The delay of mitosis*	31
	2.4.2 *The action of adrenalin*	32
	2.4.3 *The control of cell ageing*	34
	2.4.4 *The nature of the G2 chalone*	35
2.5	*Summary and conclusions*	35
	2.5.1 *The messenger molecules: hypothesis*	35
	2.5.2 *Epidermal cell competence*	38
	2.5.3 *Mitotic control: the probability theory*	38

According to the original theory of a simple negative feedback mechanism dependent on the action of a tissue-specific antimitotic messenger molecule, it was naturally to be expected that only one such substance would be needed in each tissue to ensure cellular homeostasis. Mitosis would develop automatically whenever the chalone concentration became inadequate, and would cease whenever the chalone concentration exceeded a certain critical level, which would occur whenever the cell mass was great enough. With the progress of chalone research, particularly on the epidermis, this simple concept quickly became untenable.

In the first place it had long been known, both in the embryo and in the adult, that mitotic activity in the various types of epithelium, including the epidermis, is normally stimulated by a chemical messenger, called the mesenchymal factor, which originates in the adjacent connective tissue (Wessells, 1977). Thus mitosis does not simply occur in any group of cells whenever the chalone concentration is too low. The mitosis operon must be triggered into activity by the mesenchymal factor, and it is evidently for this same reason that a 'serum factor' is needed to promote mitosis in cells *in vitro*.

Secondly, it was soon discovered that whenever the epidermal mitotic rate increases to make good the cell loss, as it does for instance after epidermal damage, this is accompanied by an equal increase in the rate of post-mitotic cell ageing and death (Bullough, 1972). Cell loss continues to balance cell gain and in such a situation a simple negative feedback mechanism could not operate.

Thirdly, the situation was complicated by the discovery that the epidermis contains not one but two evidently quite separate chalones. The first chalone to be discovered, using the colcemid technique, was found to inhibit at a point immediately before the beginning of the visible mitosis, that is after the chromosomal DNA had already been duplicated, whereas theory demanded that a chalone should inhibit at a much earlier point before the beginning of the whole mitotic process. Shortly afterwards, the second chalone, discovered by the tritiated thymidine technique, was found to inhibit at precisely this earlier point (Marks, 1971, 1976).

Thus in considering the nature of the mechanism that ensures epidermal cellular and functional homeostasis there are at least three different messenger molecules to be considered, and perhaps significantly, there are also three main cellular functions that need to be controlled. The first of these is the rate of new cell production, the second is the process of keratinization, and the third is the rate of post-mitotic cell ageing and death.

2.1 Cell gain and cell loss

Before considering the actions of the three messenger molecules it is necessary to define more closely the events that occur as a tissue cell passes from one mitosis to the next, and also as a cell passes from its final mitosis through tissue function to death. For this second and final stage of differentiation the term maturation is appropriate.

2.1.1 *The mitotic cycle*

The events that occur as a cell passes from mitosis to mitosis are still poorly understood (see Prescott, 1976), but a number of successive phases of activity have been recognized (Bullough, 1965). Of these the most obvious are the phases of DNA duplication (the S phase) and of mitosis (the M phase). Between the end of the M phase and the beginning of the S phase of the next mitosis is a relatively long gap (the interphase or G_1 phase), while between the end of the S phase and the beginning of the M phase is a relatively short gap (the antephase or G_2 phase). Although little is still known of the processes occurring during the two gaps it is possible to recognize the following subdivisions of the cycle.

The interphase or G_1 *phase*, which is divisible into at least two sub-phases:
  (1) *The apophase* or G_{1a} *phase*. This follows the completion of the visible mitosis and may perhaps be more properly regarded as the final phase of the mitotic cycle when each daughter cell re-establishes its normal structure and function. In epidermis it may typically last for perhaps 24 hours.

  (2) *The dichophase* or G_{1b} *phase*. This is the time when the cell 'rests' prior to the critical decision whether once more to enter the mitotic process, or whether instead to begin the post-mitotic tissue-specific syntheses which in epidermis lead to keratin production and cell death. In epidermis the duration of this phase varies widely from a few days when the mitotic rate is high to a few months when the mitotic rate is low; in normal epidermis it may last for some 2 or 3 weeks. It is on the cells in this phase that one of the chalones acts to inhibit entry into the mitotic process, and this chalone is therefore called the G_1 chalone.

The prosphase, which is the period of preparation for mitosis, and which is divisible into at least three sub-phases:
  (1) *The G_{1c} phase.* This is the period between the triggering into action of

the mitosis genes and the beginning of DNA duplication. In some tissue (e.g. progranulocytes) it is so short as to be almost non-existent; in epidermis it may last for perhaps 2 hours. It is character-ized by accelerated mRNA and enzyme synthesis, a process that includes the production of DNA polymerase and of the enzymes that promote histone synthesis (Bullough, 1965).

(2) *The phase of DNA synthesis,* or *S phase.* During this time the nuclear DNA is duplicated, a process that in epidermis may require some 10–15 hours. Owing to the active DNA synthesis the rate of mRNA synthesis may slacken. However, since the genes synthesize DNA sequentially and not simultaneously, there are probably always considerable numbers of genes available for mRNA synthesis.

(3) *The antephase,* or *G_2 phase.* In epidermis this period, according to circumstances, may last from one to several hours (Bullough and Laurence, 1966) during which the final syntheses for mitosis occur. These evidently include the production of those specialized molecules which, at the beginning of mitosis, aggregate to form the spindle apparatus. They certainly include the establishment of an energy store sufficient to allow the subsequent mitosis to pass to completion even in the absence of oxygen, or in the presence of respiratory inhibitors, or after the death of the animal (Bullough, 1950, 1952). In other words this is the time when the evidently heavy energy debt of the actual cell division is paid in advance. It is also during or at the end of this G_2 phase that the other epidermal chalone acts to inhibit entry into mitosis. This chalone is therefore called the G_2 chalone.

The mitosis phase, or *M phase,* occurs when the duplicated chromosomes condense and separate, and when the cell divides into two. In epidermis this takes from about one to several hours depending on the circum-stances (Bullough and Laurence, 1966). During this phase the synthesis of mRNA and of protein is severely depressed, which is evidently the con-sequence of the breakdown of the nuclear structure, the condensation of the chromosomes and the disruption of the endoplasmic reticulum. This degree of cellular disorder emphasizes the importance of the preliminary syntheses in the G_2 phase.

The general conclusion is that the mitotic process begins with a period of intense synthesis before and at the beginning of the S phase followed by a second period of intense synthesis during the G_2 phase. The G_1 chalone acts to inhibit the trigger (the stimulus of the mesenchymal factor) that sets the first period in motion; that is it acts to inhibit the transition from G_{1b} to G_{1c}. The G_2 chalone, in contrast, seems not to prevent but only to slow completion of the G_2 phase and to delay the transition from G_2 to M.

It must also be mentioned that in addition to these phases of the mitotic cycle, it is widely believed that certain interphase or G_{1b} phase cells may leave the cycle to enter a long-term resting phase that has been called the G_0 phase (Epifanova and Terskikh, 1969). However, there are alternative possibilities, one being that this phase is nothing more than an extended G_{1b} phase which, when the chalone effectiveness is high, may be of very long duration. Certainly there is at the moment no clear evidence for the existence of such a special class of resting cell as typified by some distinct physiological state.

2.1.2 *The ageing pathway*

In normal epidermis the mitotic cycle cells are usually in the basal layer, but in epidermis that is thickened, whether naturally as in the foot-pads of the mouse or pathologically in hyperplasia, they may extend into the lowest supra-basal cell layers. The post-mitotic cells typically comprise the supra-basal and distal cell layers.

When a cell moves distally and becomes post-mitotic it not only begins to synthesize keratin but also acquires a limited life expectancy. This process of post-mitotic cell ageing is seen in all tissues; in some it proceeds quickly while in others it is extremely slow. However, fast or slow, it never ceases and it is clear that the ageing mechanism, whatever its nature, must be a non-tissue-specific process like the mitotic cycle.

In all tissues the post-mitotic cells pass through the same sequence of phases as they traverse the ageing pathway to their death. These phases have been identified (Bullough, 1965) as follows.

(1) *The immature phase,* or *A1 phase*, in which the early post-mitotic cells begin their preparations for tissue function. The mitosis operon has been switched off and the tissue operon activated (Bernstein *et al*, 1977). In epidermis such cells are usually immediately supra-basal but when the mitotic rate is low the first steps in keratin synthesis may be detected in some of the basal cells (Christophers, 1971). One characteristic of these A1 phase cells is that, after epidermal damage,

the mitosis operon may be reactivated (and the tissue genes consequently inhibited) so that the cells may re-enter the mitotic process (see 3.1.4).

(2) *The mature phase*, or *functional phase*, or *A2 phase*, in which the tissue-specific syntheses are fully established, and in which the mitosis operon is firmly closed. However, the epidermis is peculiar in that its cells are not fully functional until after their death, when they form the protective squames of the stratum corneum. Thus the A2 phase of the epidermal cells is atypical; it is represented by the outer cells of the stratum spinosum in which keratin synthesis is active.

(3) *The dying phase*, or *D phase*, in which all gene activity has ceased and in which the nucleus has begun to degenerate. During this final phase protein synthesis continues under the control of preformed messenger RNA and enzymes, as described for instance in the epidermal stratum granulosum as well as in lens cells, feather cells and mature non-nucleated erythrocytes (Bishop *et al*, 1961; Humphreys *et al*, 1964). The nucleus breaks down and may disappear completely during this terminal period (Scott and Bell, 1964).

In summary, the sequence of events during cell ageing involves, first, the progressive loss of ability of the mitosis genes to be reactivated, so that at some time at the beginning of the mature phase it is lost altogether; second, the waxing and ultimately the waning of the activity of the tissue genes; and third, the final irrevocable closure of the whole genome.

It is curious how, with so much active research devoted in the name of cell kinetics to the manner and rate of cell production, so little attention has been paid to this complementary process, the manner and rate of post-mitotic cell ageing and death. Indeed there has been little general realization that cell ageing is a closely controlled active process, although this is clearly shown by the well-established fact that it accelerates or decelerates in step with the mitotic rate.

The nature of the mechanism that drives the ageing process is still a complete mystery, but it is clear that the rate at which ageing proceeds, like the rate at which mitosis proceeds, is tissue-specific. The period of cell ageing is, of course, the period of tissue-specific cell function, but evidence from tumours suggests that the two processes may perhaps be separable. Like mitosis, post-mitotic cell ageing is seen in carcinomata whose cells show no signs of tissue-specific syntheses. The question thus arises whether, also like mitosis, cell ageing may be gene controlled by what could be called an ageing operon, the rate of ageing then being determined by the degree of activity of the genes.

2.2 The mesenchymal factor

The conclusion is that epidermal cells show at least three phases of activity which are closely interrelated, which are evidently gene controlled, and which lead respectively to new cell production, to keratinization and to cell ageing; and that these cells are influenced by at least three different messenger molecules which may act as triggers for gene-directed syntheses. The first of these substances is the mesenchymal factor.

The existence of this chemical messenger, emanating from the connective tissues and stimulating mitosis in the closely adjacent epithelial cells, was first indicated by experiments with embryonic tissues and organs, but later its presence and action in adult tissues and organs was also confirmed. The extensive literature has been well reviewed by Wessells (1977). In the embryo the evidence comes from a wide range of tissues, while in the adult it comes especially from studies of skin and bladder. In all cases the general conclusion is the same, namely that in epithelial tissues mitotic activity is seen only in cells that are in actual contact with or are closely adjacent to the connective tissue.

2.2.1 *Mitotic stimulation*

The basic observation from studies of embryonic and adult skin from both birds and mammals is that, when the epidermis is separated from the dermis and kept *in vitro*, basal mitotic activity ceases and ultimately all the cells die, sometimes after keratinization has progressed down to the basal cells (McLoughlin, 1961; Wessells, 1962; Briggaman and Wheeler, 1968, 1971). Similarly, if epidermal cells are kept in a suspension culture, they first cease mitotic activity and then they flatten and keratinize (Green, 1977).

In the presence of the dermis this does not happen and the epidermal cells continue to be replaced by basal mitosis. Furthermore, if separated epidermis is replaced on to dermis, basal mitosis is resumed; this also happens if the dermis is inverted or if certain other types of connective tissue are substituted; and if dissociated epidermal cells are inoculated into the dermis they form cysts with the dermis-adjacent peripheral cells being mitotic and the inner cells cornifying to produce a central keratin mass. Basal mitotic activity and supra-basal keratinization can also be induced if separated epidermis is placed on collagen gels containing fibro-blasts, or the supernatant from fibroblast cultures (Karasek and Charlton, 1971), or fresh serum, or embryo extract.

Particularly significant is the work of Levine *et al* (1973), who used mesenchymal factor covalently bound to sepharose beads. They found that dissociated embryonic pancreatic cells will attach to such beads, and will then show active DNA synthesis and cell division at con-centrations of mesenchymal factor that are far lower than would be needed if the factor were merely dissolved in the culture medium. They concluded that the mesenchymal factor may normally exist as an aggregate or complex in the intercellular spaces beneath or around the basal epithelial cells, that it does not penetrate into the more distal cell layers in which no mitosis is seen, and that its stimulus to mitosis may be initiated by a cell surface action (cf. G_1 chalone action; 2.3.1).

In an established epidermis it is, of course, only the basal and immediately supra-basal cells that are competent to respond to the stimulus of the mesenchymal factor. If the epidermis is placed upside down on the dermis the originally distal cells, in which the mitosis operon is already permanently closed, cannot be induced to divide.

The main conclusion is that the mesenchymal factor, which is evidently synthesized by the dermal cells, is basically responsible for the stratific-ation of the epidermis. Mitosis in the basal layer leads to cell extrusion into the distal layers, where the stimulus to mitosis is lacking, and where the cells are free to respond to the alternative stimulus to maturation and keratinization. A similar situation exists in most if not all epithelial tissues (see Chapter 6) from their origin in the embryo to their stable state in the adult.

2.2.2 *The nature of the mesenchymal factor*

The chemical nature of the mesenchymal factor is still unknown but some clues do exist. Ronzio and Rutter (1973) have found that its action is associated with a glycoprotein of about 60 000 daltons. Similarly, Igarashi and Yaoi (1975) have identified a mitosis-stimulating glycoprotein of about 50 000 daltons, which was obtained from the 'microexudate carpet' of macromolecules laid down by fibroblasts on a glass surface. Like the mesenchymal factor this substance was considered to act on the cell surface.

Also possibly relevant is the identification of a mitosis-stimulating serum glycoprotein composed of two identical dimers each of about 60 000 daltons (Houck and Cheng, 1973). It is, of course, well known that fresh serum, like embryo extract, contains a substance that stimulates epithelial mitosis *in vitro* and this could be the mesenchymal factor entering the blood from the many connective tissues of the body.

It is also possible that the mesenchymal factor is a small molecule which, because of its tendency to aggregation with larger molecules, gives the impression of being a relatively large glycoprotein. It is now believed that many of the G_1 chalones aggregate in this way and that they too may be bound in the intercellular matrix to act on the cell surface.

If the mesenchymal factor does act on the cell surface through specific receptor sites, this would imply that a secondary messenger system must exist within the cell to transmit the stimulus to the nucleus, and so to the genes of the mitosis operon. There is no information on the nature of this secondary system.

2.3   The epidermal G_1 chalone

The other two chemical messengers known to operate within the epidermis are the G_1 and G_2 chalones. Unfortunately, the evidence so far available is inadequate for a proper understanding of their precise rôles in epidermal cellular homeostasis, and it is probable that such information will not be obtainable until both substances become available in pure form. The better known is the G_1 chalone, which on quite inadequate grounds is often considered to be the more important, and thus to deserve the name of the 'chalone proper'. Certainly its one known action is important, namely to inhibit tissue-specifically the epidermal cells from entering the mitotic process.

2.3.1   *Mitotic inhibition*

The control of DNA-dependent DNA synthesis at the beginning of the mitotic process evidently operates in a similar way to the control of DNA-dependent RNA synthesis; it is an epigenetic event initiated or inhibited by one or more specific messenger molecules. The mitotic process, sometimes taking as long as a day to be completed, must depend on the appearance in proper sequence of an array of specialized enzymes synthesized in response to the sequential activation of the genes of the mitosis operon.

This mitotic process, complex as it is, is an all-or-none reaction. Once a cell has taken the first step towards DNA synthesis, that cell will always complete its division. Thus the G1 chalone, directly or indirectly, must inhibit a trigger reaction. Once the trigger has been pulled in response to the mesenchymal factor, the mitotic process cannot be stopped by the chalone, although there is some evidence that it may be slowed.

The presence of an epidermal G1 chalone in crude extracts of mammalian skin was first reported by Hennings *et al* (1969), and the later partial purification of such extracts confirmed that it is a different substance from the G2 chalone (Elgjo *et al*, 1971, 1972; Marks, 1971, 1973). This evidence for the existence of two separate epidermal chalones was then strengthened by the finding that their peak concentrations lie in different epidermal cell layers (Elgjo *et al*, 1972). The G1 chalone is evidently in highest concentration in the distal cell layers, whence it does not diffuse into the dermis in assayable amounts, while the G2 chalone is in highest concentration in the basal cell layer, whence it diffuses both inwards into the dermis and outwards into the epidermis.

The distribution pattern of the G1 chalone helps to support the argument that this chalone is synthesized in the distal epidermal cells, and that its primary function may be not simply to inhibit mitosis but to promote post-mitotic maturation in those cells that leave the basal layer. The inhibition of the mitotic process may then be the automatic outcome of the tissue-specific stimulus to keratin synthesis. On this argument, the maturation of the epidermal cells could be a self-promoting process.

The evident confinement of the G1 chalone action to the supra-basal cell layers suggests that its movement may be sluggish and that, like the mesenchymal factor, it may aggregate with glycoproteins (see 2.3.5) in the intercellular spaces, where it may act on receptor sites on the cell surface. It is well known that cell surface proteins, which act to limit mitotic activity and which could be the chalone receptors, can be stripped from the cell surface by trypsin treatment so that the cells then burst into mitotic activity (Carney and Cunningham, 1977).

The few certain facts about the epidermal G1 chalone action in inhibiting the onset of mitosis are that it is tissue-specific (Nome, 1975), non-species-specific, non-toxic even in very high concentrations, and reversible. Both the epidermal G1 and G2 chalones are present and active, not only in epidermis proper but also in the oral and oesophageal epithelia. However, any suggestion that these chalones may be typical of keratinizing epithelia in general is counteracted by the fact that they can also be extracted from the trachea and the bronchioles, which do not keratinize except in pathological conditions. Whether they are also present in the epithelia of the salivary gland ducts, the bladder, and the urethra, all of which may also keratinize in abnormal circumstances, is not known.

Evidence of this kind is beginning to indicate that the tissue-specificity of chalone action may not necessarily be related to the usual tissues as defined by a histologist. Clearly this does not alter the basic concept of the tissue-specificity of chalone action; it merely alters the concept of a tissue in the chalone context (10.2.1).

2.3.2 *The action of the glucocorticoid hormone*

That the adrenal hormones are in some way involved in the mechanism of epidermal mitotic activity was first established through studies of the diurnal mitotic cycle in mouse epidermis (Bullough, 1948; Bullough and Laurence, 1961). It is now well known that in many mammalian tissues, including epidermis, the numbers of mitoses are low when the animal is awake and stressed and high when it is asleep and relaxed. It is also well known that injections of a glucocorticoid hormone inhibit mitotic activity by preventing the intermitotic G_{1b} cells from entering the phase of DNA duplication. Thus a glucocorticoid hormone inhibits the mitotic cycle at the same point as does the G_1 chalone (Marks, 1976). Conversely, it has been shown that adrenalectomy results in increased numbers of epidermal cells entering the mitotic process, and that a return towards normality can be achieved by glucocorticoid hormone injections (Desser-Wiest, 1974).

However, Marks (1976) has indicated that the epidermal G_1 chalone and the glucocorticoid hormone do not act through the same channel. It has already been suggested that the G_1 chalone, present in the inter-cellular spaces, may act at receptor sites on the outer cell membrane. In contrast it is known that the glucocorticoid hormones, like the other steroid hormones, act at intracellular cytoplasmic receptor sites, the receptor–steroid complex then entering the nucleus and binding with specific acceptor sites on the chromatin. It is possibly at this point that the glucocorticoid hormone mechanism may impinge on the secondary messenger system that transmits the surface chalone action to the nucleus.

It is also possible that the cyclic AMP–GMP system is in some way involved since it is widely believed that this system may play a critical rôle in mitotic control (Halprin, 1976). The glucocorticoid hormone action may have a potentiating or 'permissive' action on cyclic AMP mediated processes, and it has even been suggested that the actions of a glucocorticoid hormone and of cyclic AMP could have some distal steps in common (Gehring and Coffino, 1977).

However, these are speculations and the only certain conclusion is that glucocorticoid hormone action in some way impinges on and co-operates in that chain of events that lies between the initial G_1 chalone action and the ultimate nuclear response.

One other potentially important point does however arise. Studies of the actions of a glucocorticoid hormone on cells *in vitro* have shown a stimulus to maturation. Thus Endo *et al* (1977) have found that 13-day-old chick epidermis *in vitro* undergoes dramatic keratinization in the presence of minimal concentrations of hydrocortisone, this action being related to the induction of long-lived messenger RNA leading to the synthesis of stable tonofilamentous proteins. In a normal chick embryo adrenocortical function begins at 13–15 days, and is followed by epidermal keratinization at 18 days. The conclusion reached was that the completion of epidermal keratinization is corticosteroid-dependent.

This is in accord with the suggestion that the primary action of the epidermal G_1 chalone, supported by the glucocorticoid hormone, is not so much to inhibit the mitotic process as to promote epidermal maturation with keratinization.

2.3.3   *The control of cell maturation*

Although the epidermal G_1 chalone, like all other chalones, is tissue-specific in its antimitotic action, the process of mitosis is obviously non-tissue-specific. To meet this difficulty it has been suggested that the chalone may act by cleaving to tissue-specific receptor sites on the outer cell membrane. Through these receptor sites it may then either activate a non-tissue-specific secondary messenger system which directly inhibits the mitosis genes, or it may promote tissue-specific cellular maturation which then leads indirectly to mitotic inhibition (Chopra, 1978).

In the absence of concrete evidence both these suggestions must remain as possibilities. Certainly it seems that a negative feedback link does exist between the tissue operon and the mitosis operon; so long as the tissue genes are activated the mitosis genes must remain silent.

The activation of cell maturation, unlike the activation of the mitotic process, is not at first an all-or-none reaction. After any form of tissue damage or loss, when the chalone concentration falls, the younger post-mitotic cells can revert to mitosis once more (see 4.4.2). It follows that if the G_1 chalone activates the tissue operon, a continued effective concentration of this chalone is necessary to maintain the stimulus to maturation, at least during the earlier phase when the mitosis operon remains capable of reactivation.

This concept would fit with the suggestion that not only is an inductive stimulus needed for the original embryonic differentiation of a tissue but that this stimulus, or another similar stimulus, must be maintained throughout life to ensure the continued existence of that tissue.

2.3.4 *Fetal and neonatal epidermis*

In the mouse, from which most information comes, full epidermal differentiation is not completed until some 2 weeks after birth (Gibbs, 1941). During these 2 weeks the epidermal cells remain multipotential, and they give rise to eccrine glands, to hair follicles with sebaceous glands, and to the epidermis proper, all of which then develop their own specific chalone control systems.

It has been found that, during this phase of active differentiation of epidermal derivative tissues, epidermal mitotic activity cannot be inhibited by the G_1 epidermal chalone, and also that epidermal mitotic activity cannot be stimulated by chemical or physical damage (Bertsch and Marks, 1974). Nevertheless, both the G_1 and G_2 chalones are present, and indeed neonatal epidermis can be used as a good source for chalone extraction. It has been shown that in both embryonic chick and fetal mouse the G_1 chalone is detectable from the time that keratinization is first seen, which has led Bertsch and Marks to conclude that 'the production of the G_1 chalone may be a consequence of the expression of tissue function which would be consistent with its tissue-specificity'.

After the 2 week postnatal period there is a marked change in the appearance and behaviour of mouse epidermis. In a newborn mouse the epidermis is relatively thick and its mitotic activity is relatively low. As it expands in area over the rapidly growing body it shrinks to about a third of its original thickness. Then from about day 14 the mitotic rate rises sharply, epidermal thickness keeps pace with the expanding body surface, and the typical adult epidermal structure is attained. Simultaneously, the epidermal cells lose their ability to give rise to other types of tissue, they begin to respond by mitotic inhibition to the two chalones, and they show mitotic stimulation after damage. Epidermal differentiation is complete.

There are two possible theories to explain the failure of the neonatal epidermal cells to respond to the epidermal chalones. The first is that other essential factors, such as the glucocorticoid hormone (2.3.2), may not yet be present in adequate amounts. The second is that neonatal epidermis may possess its own specific control mechanism. In this connection it is well established that the multipotential stem cells of the haemopoietic system do have their own control system, and do not respond to the chalones of the erythrocytic and granulocytic tissues that are derived from them (8.2.1). The neonatal epidermal cells are also multipotential stem cells in that they can give rise to several different types of tissue, each of which has its own individual chalone control system.

Evidently as a tissue differentiates into existence it simultaneously acquires a chalone control system and the competence to respond to it; the epidermis of the mouse is not finally and fully differentiated until 2 weeks after birth.

2.3.5 *The nature of the G1 chalone*

The main attempts to understand the chemical nature of the G1 chalone began using aqueous extracts of powdered pig skin (Marks, 1976). In such extracts the chalone has proved to be remarkably stable; it is not destroyed by boiling, and it is resistant to a range of enzymes including trypsin, pronase, and ribonuclease. So stable is it that *in vivo* it may prove to have a long life. A purification of almost a million-fold has been achieved, and although the final fraction was still heterogeneous, containing RNA, several mucopolysaccharides, and some protein-like material, Marks has concluded that the active principle may be a glycoprotein, and that its resistance to proteases may be due to the protective action of its carbohydrate components.

However, it is also clear that the chalone aggregates strongly with other molecules, and that this has led to overestimates of its molecular weight. After molecular dis-aggregation a figure of a little less than 10 000 daltons has been suggested, but it is probable that this is still much too high (Patt and Houck, 1980). A comparison with the other G1 chalones so far studied (see 6.1.3) suggests that these substances could prove to be small polypeptides (8.2.3), some of which have a strong tendency to aggregate with mucopolysaccharides. As already noted, mucopolysaccharides are evidently major components of the epidermal intercellular cement, which may, therefore, bind both the mitosis-controlling messenger molecules, the mesenchymal factor and the G1 chalone.

Using the most highly purified preparation of the epidermal G1 chalone so far available, Marks (1975) found that a dose of 0.05–0.1 μg injected intraperitoneally into a mouse was sufficient to induce a 50–70% inhibition of basal cell entry into mitosis. Since this low dose can certainly be reduced after further purification, it is clear that the G_1 chalone must act at a cell concentration similar to that of a hormone in its target tissue. This emphasizes the main difficulty in chalone purification; to obtain a fraction of a milligram of pure epidermal chalone it is necessary to start with many kilograms of defatted lyophilized skin powder, which represents the whole skins of many full-grown pigs.

2.4 The epidermal G_2 chalone

This was the first tissue-specific mitotic inhibitor to be discovered and to be named a chalone (Bullough, 1962). Its presence was detected in extracts of mouse skin and pig skin (Bullough and Laurence, 1964a, Bullough et al, 1967), which probably also contained the still unknown G_1 chalone. However, in spite of its earlier discovery, less is known of its rôle in epidermal homeostasis and of its chemical structure than is the case with the G_1 chalone.

2.4.1 The delay of mitosis

The known facts about the epidermal G_2 chalone are that it delays but does not prevent the passage of a G_2 or antephase cell into mitosis; that it is tissue-specific but non-species-specific in its action, which is non-toxic and reversible. Also its action is related to the presence of adrenalin. In the absence of this hormone the G_2 chalone action is severely weakened, while in normal epidermis, with some adrenalin always present, the power of the G_2 chalone is proportional to the adrenalin concentration (Bullough and Laurence, 1964b; see 2.4.2). It is also known that, with adrenalin, the G_2 chalone slows the passage of a cell through mitosis, which is probably the consequence of its earlier action in the G_2 phase.

Like the epidermal G_1 chalone, the G_2 chalone is present and active not only in the epidermis proper but also in the epithelia of the palate, oesophagus, and lens of the eye; it is present in the trachea and the bronchioles, but it has no action on the sebaceous and sweat glands, on intraepidermal melanocytes, on dermis or on other skin tissues (Nome, 1975).

At the moment, it is difficult to see any significance in the delay induced by the chalone on the entry into mitosis of cells that are already prepared and committed to divide. One theory, proposed by Gelfant and Candelas (1972), and contradicted by Sauerborn et al (1978), is that it creates a reserve of arrested G_2 phase cells, which are ready to move immediately into mitosis after skin damage. Certainly a small number of arrested G_2 phase cells is present in the epidermis when an animal is awake with a relatively high blood adrenalin concentration, but all these cells are released to pass through mitosis when the adrenalin concentration falls during rest or sleep (Bullough and Laurence, 1961). No constant reserve of arrested G_2 phase cells exists.

Another theory, proposed by Elgjo (1975), is that the G_2 chalone may act as a 'differentiation factor', which so alters the programme of syntheses within the G_2 phase cells that, when they complete their division, they are conditioned to begin keratin synthesis. This idea derives from the old hypothesis that any inductive event in tissue differentiation occurs during a preceding, or quantal, mitosis. In epidermis this is certainly not true: both the daughter cells of any mitosis tend to remain in the basal mitotic cell layer (4.2.1), and the induction of keratin synthesis begins in those other cells that are fortuitously displaced into a supra-basal position. Cellular maturation is determined by the supra-basal cellular environment.

There is a third possibility that so far remains unexplored; namely that the main inhibitory action of the epidermal G_2 chalone may relate not to the mitotic process but to the ageing process of the post-mitotic cells. This possibility is discussed below in 2.4.3.

2.4.2 *The action of adrenalin*

While the glucocorticoid hormone acts to support epidermal G_1 chalone action, adrenalin acts to support epidermal G_2 chalone action. In the presence of the G_2 chalone, adrenalin powerfully inhibits the entry of an epidermal G_2 phase cell into mitosis, and thus is primarily responsible for the well-known diurnal mitotic cycle. In man the rate of adrenalin secretion is some 3–10 times higher when awake than when asleep, and this is almost enough to prevent the onset of epidermal mitosis. During subsequent sleep all the arrested cells enter mitosis together (Bullough, 1962). After adrenalectomy the diurnal mitotic cycle disappears (Bullough and Laurence, 1961).

It seems that neither adrenalin alone nor epidermal G_2 chalone alone is adequate to prevent the onset of mitosis; both must be present for a strong inhibition to develop (Bullough and Laurence, 1964b). It follows that in damaged epidermis with a low chalone concentration and a high mitotic rate the sensitivity to adrenalin inhibition is greatly reduced, and similarly in tissues with a naturally high mitotic rate, such as duodenal mucosa, seminiferous tubules and active hair bulbs, all of which are thought to have a relatively low chalone content, there is little or no diurnal mitotic cycle.

Adrenalin also slows the rate of completion of mitosis in any cell that manages to pass the G_2 blockage point, but curiously it accelerates the completion of any mitosis that is already in existence when the adrenalin concentration rises (Bullough and Laurence, 1966). The implication is that this slowing action is not direct but is the result of the previous influence of the adrenalin on the G_2 phase. The same slowing action can be obtained by raising the G_2 chalone concentration, and the conclusion is that the inhibiting action of both these substances is confined to the G_2 phase.

A theory has been developed that the antimitotic action of adrenalin operates through changes induced in the intracellular cyclic AMP or GMP concentrations (Halprin, 1976; Green, 1978). Adrenalin does not enter the target cell, but by binding to receptor sites on the cell membrane it interacts, directly or indirectly, with the cyclic AMP–GMP system, which may thus form part of a secondary messenger system. It has been shown that cyclic AMP is present in epidermal cells, and that this substance can inhibit epidermal mitosis in the G_2 phase (Marks and Rebien, 1972). However, the metabolic pathways of which the cyclic AMP–GMP system forms a part are not specific to the process of mitotic control, and no connection has yet been found between this system and the action of the G_2 chalone.

One final point concerns the influence of the glucocorticoid hormone on adrenalin action. Iizuka *et al* (1980) have found that, when adrenalin causes an increase in the cyclic AMP content of the epidermis, hydrocortisone (which has no direct effect on cyclic AMP) 'protects the adrenalin-adenylate cyclase system of the skin' and so prolongs the adrenalin effect. In exactly the same way, Bullough and Laurence (1968a) have shown that, when adrenalin causes an inhibition of epidermal mitosis, hydrocortisone (which has no direct effect on the G_2 phase) acts to prolong the adrenalin effect.

The general conclusion must be that both the stress hormones co-operate in some way to condition the epidermal cells to respond to the G_2 chalone.

2.4.3 *The control of cell ageing*

As already explained, the rate of post-mitotic cell ageing is tissue-specific-ally controlled in such a way that the faster (or slower) the rate of cell gain by mitosis, the faster (or slower) the rate of cell loss by ageing and death. The slender evidence that the G_2 chalone system is linked to the control of cell ageing is as follows:

(1) Any form of stress that results in an increased rate of secretion of the two stress hormones also results in a slower rate of post-mitotic cell ageing and death (Bullough and Ebling, 1952). Conversely, adrenalectomy, which results in an increased rate of cell production, does not disturb the balance between cell gain and cell loss, and must, therefore, also result in an increased rate of cell ageing and death. The rate of post-mitotic cell ageing is inversely proportional to the stress hormone concentration.

(2) More important is the evidence from the erythrocyte and gran-ulocyte systems, each of which possesses its own specific G_1 chalone that slows the rate of new cell production in co-operation with the glucocorticoid hormones (8.2.2; 8.2.4). Both these tissues are peculiar in that they do not possess a G_2 chalone, that the ageing rate of their post-mitotic cells is not related to the mitotic rate, and that neither responds to changes in the adrenalin concentration. This negative evidence supports the view that in those tissues that do possess a G_2 chalone, this chalone together with adrenalin may play a part in the control of cell ageing and death.

On this theory, the basal epidermal cells, which contain the highest G_2 chalone concentration, do not age; only the supra-basal cells in which the G_2 chalone concentration is lower, can begin the ageing process. Also on this theory, in stressful conditions the extra glucocorticoid hormone strengthens the G_1 chalone action and the extra adrenalin strengthens the G_2 chalone action so that cell gain and cell loss are equally reduced, while conversely after epidermal damage, when both chalones are reduced in con-centration and effectiveness, cell gain and cell loss are equally increased.

One problem that arises concerns the way in which a G_2 chalone mechanism can control tissue-specifically the non-tissue-specific process of post-mitotic cell ageing. As with the similar problem of mitotic control, one solution could be for the chalone to cleave to tissue-specific receptor sites on the outer cell membrane, and thus to activate a non-tissue-specific secondary messenger system within the cell, a system that may only be fully functional in the presence of an adequate concentration of adrenalin.

2.4.4 The nature of the G_2 chalone

The first attempt to learn something of the chemistry of this chalone was by Bullough *et al* (1964) using extracts of mouse epidermis. The active principle proved to be water-soluble, non-dialysable, precipitated between 70% and 80% ethanol and destroyed by boiling, and it was concluded that the G_2 chalone is probably a protein. The next attempt by Hondius Boldingh and Laurence (1968), using extracts of pig skin, confirmed the original conclusions and suggested a molecular weight of about 30 000–40 000. Still further purification by Isaaksson-Forser *et al* (1977) has indicated that this chalone may be a glycoprotein of about 22 000 daltons.

However, all the material so far studied has been very impure, and the suggested molecular weights have little significance. Marks (1976) has emphasized that the G_2 chalone molecules may be aggregated to larger molecules, as happens with the G_1 chalone, and that the apparent molecular weights of both the epidermal chalones may, therefore, vary according to the extraction and purification techniques used. There is some evidence that, when the aggregated molecules have been stripped away, the epidermal G_2 chalone may prove to have a low molecular weight similar to that proposed for the various G_1 chalones (6.1.3).

Thus the chemical nature of the epidermal G_2 chalone remains obscure, and since no attempts have yet been made to characterize any of the other G_2 chalones, no useful comparisons can be made.

2.5 Summary and conclusions

The epidermal cells, like the cells of all other tissues, show at least two alternative states of differentiation: the mitotic state and the state of tissue function with cell ageing. The evidence also suggests that these last two, function and ageing, may be separable, at least in pathological conditions, and if this is true the epidermal cells may be capable of three separate activities, each of which may be initiated or controlled by a specific messenger molecule. There are also three known messenger molecules in the epidermis of which one, the mesenchymal factor, triggers into action the mitosis operon. The other two, the G_1 and G_2 chalones, may perhaps prove to control keratin synthesis and post-mitotic cell ageing respectively.

2.5.1 The messenger molecules: hypothesis

As a working hypothesis it is here assumed that epidermal cellular homeo-stasis, that is the balanced relationship between the rates of cell gain, cell

35

function, and cell loss, may be controlled by the three known messenger molecules, as follows:

The mesenchymal factor, presumably synthesized by the dermal cells, stimulates the rate of cell production by increasing the probability that the basal cells will enter the mitotic process (see 2.5.3), the result being a shortening of the intermitotic interval. Being bound to the intercellular matrix, it penetrates, in effective concentration, only a short distance into the epidermis. The limit to its stimulus, in conditions of extreme hyperplasia, is set by the minimum time needed for the completion of the sequence of syntheses that constitute the mitotic cycle. With the intermitotic, or G_{1b} interval, reduced almost to zero, the time needed for the mitotic process itself may be as little as 12 hours, thus allowing two mitoses per day as in active hair bulbs.

The epidermal G_1 chalone, found in the highest concentration in distal keratinizing epidermal cells where it is presumably synthesized, acts tissue-specifically to promote the post-mitotic state with keratin synthesis. As the chalone penetrates down into the basal cell layer it inhibits the rate of cell production by reducing the probability that the cells will enter the mitotic process. The mitotic rate is thus a function of the relative concentrations in the basal cell layer of the mesenchymal factor and the G_1 chalone. When the G_1 chalone concentration is relatively high, the mitotic rate is low, and some of the cells while still basal may begin to synthesize keratin precursors (Christophers, 1971); when the G_1 chalone concentration is relatively low, the mitotic rate is high, and the influence of the mesenchymal factor may extend into the lower supra-basal cell layers to enable these cells to continue to divide.

As the post-mitotic epidermal cells rise towards the surface the stages of keratin synthesis seen in each cell layer are closely synchronized. This could be due to the high accuracy of the timing mechanism that underlies the ageing process, but it could also be achieved by chemical communication through a system of intercellular connections which enable each cell layer to function as a continuum (Karasek and Liu, 1977).

The epidermal G_2 chalone, in highest concentration in the basal cell layer where it is evidently synthesized, acts tissue-specifically to delay, but not to prevent, part of the mitotic process. However, it is possible that its main action may be to slow the rate of ageing, and thus of keratin synthesis, in the post-mitotic cells. In these cells a timing device is activated which ensures that, after a tissue-specific interval, the mitosis operon is closed, and that after a second tissue-specific interval, the tissue operon (or keratin operon) is also closed. When the G_2 chalone action is strengthened by adrenalin the timing device is slowed down; when the G_2 chalone action is weakened after tissue damage the timing device is speeded up.

Whether or not the two epidermal chalones act in the ways suggested cannot be finally determined until highly purified preparations become available. In the meantime, using the present theory, it is interesting to consider what the epidermal cells would be expected to do if they received no instructions from their environment or if the instructions they received were unbalanced.

First, in the absence of any effective instructions, it would be expected that the cells would merely survive inertly. This may be the situation in a basal cell in which the G_1 chalone action is so powerful, as during chronic stress, that it nullifies the mitotic stimulus of the mesenchymal factor. The cell then remains inactive in what has been called the G_0 phase until the situation changes.

Second, if the epidermal cells were to receive only the basal stimulus of the mesenchymal factor, mitotic activity would continue at a high rate, and the cells that were forced distally, lacking both the G_1 chalone influence for keratinization and the G_2 chalone influence to delay ageing, would die rapidly without maturation. A situation approaching this is found in certain anaplastic epidermal carcinomata (Bullough and Laurence, 1986b, Bullough and Deol, 1971b) in which the chalone concentrations are known to be abnormally low (see 9.2.1).

Third, if the basal epidermal cells lacked the mesenchymal factor while both the chalones were present, all the cells would be expected to become post-mitotic, to keratinize and to die. This situation and this response have been described, for instance by McLoughlin (1961), in epidermis that has been isolated from its dermis.

One final point may be stressed, namely that if the G_1 chalone is the critical factor inducing the post-mitotic keratin-synthesizing stage, then it is also essential for the continuing existence of the epidermis. In this case the single most important substance synthesized in any tissue may be its G_1 chalone on which the tissue's very survival depends. On this point Gurdon (1977) has suggested that in any tissue cell the products of certain genes, accumulating in the cytoplasm, may re-enter the nucleus to stimulate the continued function of the genes, and Bullough (1965) has emphasized that not only the functional state of a tissue but the whole structure of the adult mammal may need at all times to be actively maintained against collapse into mitotic anarchy.

2.5.2 *Epidermal cell competence*

The evidence suggests that both the epidermal chalones may cleave to tissue-specific receptor sites on the cell surface, and thus may activate intracellular secondary messenger systems, which are themselves influenced by the actions of the two stress hormones. Clearly, only when these secondary systems are fully functional, and when the receptor sites are coupled to them, can a cell respond to the chalone message.

In the newborn mouse the epidermal cells are evidently not competent to respond to the adult-type epidermal G_1 chalone, although this chalone is already present. At this time the surface receptor sites may respond only to the G_1 chalone typical of fetal-type epidermis, which is a multipotential stem cell type of epithelium. It is only when the adult-type epidermis becomes fully differentiated at about 14 days of age that the coupling to the adult-type G_1 chalone occurs.

This recalls the situation in an embryo where the differentiating cells must be competent to respond before they can react to the message of the evocator molecules.

2.5.3 *Mitotic control: the probability theory*

An important contribution to the understanding of the manner of entry of a G_{1b} phase, or dichophase, cell into mitosis is that of Smith and Martin (1973) (Shields, 1978). Contrary to the older assumption that in any particular type of cell the duration of the intermitotic G_{1b} phase is fixed, they have proposed that it should be expressed as a half-life, each intermitotic cell at each moment having an equal chance 'that DNA synthesis will commence within the next minute'. On this theory it is the probability that this will happen in a basal epidermal cell that is dependent on the relative concentrations, or strengths of actions, of the mitosis-stimulating mesenchymal factor and the mitosis-inhibiting G_1 chalone.

The importance of this concept has also been shown theoretically by the computer studies of Mitrani (see Bullough and Mitrani, 1976), which took account not only of the rate of basal epidermal cell entry into the mitotic process but also of the manner and rate of basal cell extrusion into the supra-basal cell layers. Because basal cell extrusion occurs only in the immediate proximity of a mitosis (4.2.1), it was found that, if the intermitotic interval is of fixed duration, an initially unsynchronized cell population must eventually become synchronized with all mitoses occurring simultaneously, which is contrary to observation. This would not occur if the intermitotic interval includes an element of probability.

In consequence, a theory has been put forward (Bullough and Mitrani, 1976) that the fate of an intermitotic G_{1b} cell may depend on the number, or proportion, of receptor sites that are occupied by chalone molecules. If the occupation of a receptor site is unstable, so that the situation is constantly changing, entry into the mitotic process would depend on a small enough number of G_1 chalone sites, and a large enough number of mesenchymal factor sites being occupied for a long enough period for the trigger to operate. Thereafter, as has been emphasized, the whole mitotic process must proceed to completion. The converse situation would then develop in the supra-basal cell layers in which the state of post-mitosis with maturation, which for a time is reversible, would depend on a large enough number of receptor sites being constantly occupied by the G_1 chalone.

It can be added that the entry of a G_2 phase epidermal cell into mitosis is also likely to be a probability event, the probability being high during sleep and low during waking.

Chapter 3

Epidermal cellular homeostasis

3.1	*The cell gain : cell loss ratio*		41
	3.1.1	*The ratio constants*	41
	3.1.2	*Phase 1 epidermis: hypoplasia*	42
	3.1.3	*Phase 2 epidermis: hyperplasia*	44
	3.1.4	*Epidermal wound healing*	45
	3.1.5	*G₁ chalone action in hyperplasia*	46
3.2	*The effects of stress*		47
	3.2.1	*Stress and cell production*	48
	3.2.2	*Age and stress*	49
3.3	*A chalone negative feedback mechanism*		50
	3.3.1	*General chalone loss*	50
	3.3.2	*General chalone excess*	51
3.4	*Summary and conclusions*		51
	3.4.1	*The chalone control mechanism*	52
	3.4.2	*The rate ratio*	52
	3.4.3	*Post-mitotic cell ageing*	53

The general picture emerges of epidermis as a tissue whose continued function, and indeed continued existence, depends on a tissue-specific chalone mechanism which initiates and controls the synthesis of keratin, and which interacts with two non-tissue-specific mechanisms that control, respectively, cell gain by mitosis and cell loss by post-mitotic cell ageing. The chalone mechanism ensures that throughout a long adult life keratin synthesis is adequate and epidermal cellular homeostasis is maintained.

In normal circumstances the number of epidermal cells, and thus the epidermal thickness, varies little, while in the abnormal circumstances of extensive cell destruction the normal cell number is quickly restored.

Clearly the epidermal homeostatic mechanism is more complex than was envisaged in the original simple negative feedback theory. How the mechanism actually works must now be considered.

3.1   The cell gain : cell loss ratio

The close connection between the rates of cell gain and cell loss was first shown in studies of epidermis and sebaceous glands (Bullough and Ebling, 1952) and it is now clear that the same connection also exists in the other epithelial tissues (Bullough, 1965, 1967). In fact it is this compensating linkage that is basically responsible for cellular homeostasis in the epidermis and other tissues.

3.1.1   *The ratio constants*

In epidermis the rates of mitosis and of post-mitotic cell ageing are linked in such a way that:

rate of cell gain : rate of cell loss = R, a constant.

At the one extreme it has been shown that when, in experimental hypoplasia, the mitotic rate is maintained at a quarter of its normal level for several weeks, there is no change in epidermal thickness. Clearly in these circumstances the time taken for each post-mitotic cell to rise to the surface and to keratinize must have increased to about × 4 normal, that is from about 3 weeks to about 3 months. Conversely, in hyperplastic epidermis with a greatly increased mitotic rate, the time taken by the post-mitotic cells to age and die is reduced from the normal 3 weeks to only about 4 days (Scott and Ekel, 1963).

Although, by definition, the epidermis becomes thinner in hypoplasia and thicker in hyperplasia (for explanation see 3.1.3), the rate ratio, or R ratio, remains constant over a very wide range. The width of this range is indicated by the fact that, when mouse epidermis passes from hypoplasia to extreme hyperplasia, the mitotic rate may increase more than a hundred-fold while at the same time, because of the R ratio, the increase in epidermal thickness is limited to only some 2- or 3-fold (Bullough and Stolze, 1981, unpublished data).

Furthermore, over this whole range, the number of post-mitotic epidermal cells remains in proportion to the number of basal, or mitotic-cycle, cells (Bullough, 1975) so that:

number of mitotic-cycle cells : number of post-mitotic cells = N, a constant.

This implies that the normal thickness of the epidermis is determined by the value of R, and therefore also of N. The slower the rate of post-mitotic cell ageing relative to the rate of mitosis (i.e. the lower the value of R), the greater is the number of post-mitotic cells relative to the number of mitotic-cycle cells and the thicker is the epidermis. The values of R and N are evidently both epidermis-specific and species-specific; other tissues with other values have other masses, while other species with other values for the epidermis have other epidermal thicknesses.

It has been suggested that the rate of mitosis is a function of the G_1 chalone concentration, while the rate of post-mitotic cell ageing may be a function of the G_2 chalone concentration. If this is true the value of the R ratio may be determined by the ratio of the concentrations of the G_1 and G_2 chalones, and this could indeed be the basic role of the two chalones. On this theory the higher the G_2 chalone concentration the slower the rate of post-mitotic cell ageing and the thicker the epidermis. If it should prove that the G_2 chalone does not function in this way then it may be necessary to search for some other chemical messenger that does.

Two questions now arise: if the rates of cell gain and cell loss are always so perfectly balanced as the evidence indicates, how is it that a reduced mitotic rate leads to epidermal thinning (hypoplasia) and a raised mitotic rate leads to epidermal thickening (hyperplasia), and how is it that after massive cell loss due to damage or wounding the increased mitotic rate is able to create the large numbers of new cells needed to repair the damage? These questions are considered in sections 3.1.2–3.1.4.

3.1.2 *Phase 1 epidermis: hypoplasia*

Mammalian epidermis can exist in at least two distinct states, which have been termed phase 1 and phase 2 (Bullough, 1972). Phase 1 epidermis, which is typified by mouse epidermis, is characterized by its relative thinness, by a lower mitotic rate and slower post-mitotic cell ageing, and by a flattened dermo-epidermal junction. Within the limits of this phase, whether normal as in the mouse or abnormal as in man, any chronic change in the mitotic rate has no effect on epidermal thickness since it is completely offset by a similar change in the rate of post-mitotic ageing.

This is obviously important for survival because when, for any reason, this state has been reached, the epidermis does not become any thinner and so remains adequate for body protection. In human epidermal hypoplasia the post-mitotic cells that would normally have aged, keratinized and died in 2 or 3 weeks now survive for many months. It is impossible to slow the rate of new cell production to the point when the outermost cell layers and the stratum corneum are in danger of disappearing.

Curiously in these circumstances the rate of loss of the dead squames from the outer epidermal surface is also reduced. Evidently the bonds cementing the squames together do not simply break after a given period of chemical disintegration; the period of disintegration is variable. Possibly in hypoplasia, with the slower rate of keratinization, a more efficient binding is created (4.5.4), which then takes longer to break down; conversely in hyperplasia, with fast keratinization, the squames are not so tightly bound together and are more readily shed.

In the hypoplastic phase 1 type of epidermis it has been found that the basal cell layer contains not only mitotic-cycle cells but also cells which already contain keratin precursors (Christophers, 1971) and which are already in the early post-mitotic A_1 phase. When a mitotic-cycle cell divides both its daughter cells remain basal (Bullough and Laurence, 1964a), the lateral tension so created being relieved by the extrusion distally of a neighbouring A_1 phase cell (MacKenzie, 1970; Christophers, 1971).

The higher (or lower) the mitotic rate, the more (or fewer) A_1 phase cells are forced distally, and the faster (or slower) they then keratinize and die. The number of epidermal cells, and therefore the epidermal thickness, remains unchanged.

Because in this type of epidermis the basal cell layer contains both mitotic-cycle cells and early post-mitotic cells, it is necessary to modify the N ratio as follows:

number of basal cells : number of supra-basal cells = N, a constant.

Regarding the basal post-mitotic cells it must be emphasized that so long as they remain basal they do not behave as do those post-mitotic cells that have left the baseline. While they continue in dermal contact they do not begin to age, they do not proceed far with keratin synthesis and they remain small (when they become supra-basal they increase about ten-fold in volume). Until forced from the baseline by some closely adjacent mitosis they remain in a post-mitotic limbo.

It seems that in phase 1 epidermis the lower the mitotic rate the greater the proportion of the basal epidermal cells that are in this arrested post-mitotic A_1 state. Conversely, the higher the mitotic rate the lower the proportion of these A_1 phase cells until finally, at a critical mitotic rate, all the basal cells are in the mitotic cycle. At this point phase 1 ends and phase 2 begins.

3.1.3   *Phase 2 epidermis: hyperplasia*

This type of epidermis is seen, for instance, in normal human skin as well as in all forms of pathological hyperplasia. It is characterized by its relative thickness, by a higher mitotic rate and faster post-mitotic ageing, by a single or multiple basal mitotic cell layer or layers, and commonly by a folded dermo-epidermal junction. It is also characterized by the fact that any increase in the mitotic rate is accompanied by an increase in the epidermal thickness, even though it is clear that both the R and N ratios remain constant (Bullough, 1972).

Any increase in the mitotic rate depends, first, on a shortening of the mitotic cycle so that each cell divides more often, and second, on an increase in the number of cells that are involved in the mitotic cycle. Both of these effects are caused by the weaker action of the G_1 chalone and the consequent stronger action of the mesenchymal factor. The resulting stimulus both increases the probability that each G_{1b} phase cell will re-enter mitosis so that it does so earlier, and deepens the penetration of the mesenchymal factor stimulus into the epidermis so that more cells remain in the mitotic cycle.

In phase 1 epidermis the increase in the number of basal mitotic-cycle cells is at the expense of the number of basal A_1 phase cells, so that there is no increase in the number of basal cells, and consequently no increase in the number of supra-basal cells. Epidermal thickness remains unchanged. In phase 2 epidermis, however, the increase in the number of basal mitotic-cycle cells is absolute, the extra cells being accommodated in the immediate supra-basal cell layer or layers, as in hyperplastic mouse epidermis (Bullough and Deol, 1975), or on the extra baseline area created by the deeper folding of the dermo-epidermal junction, as in man (Tosti *et al*, 1959, 1969; Bullough and Stolze, 1981, unpublished data), or in both these ways. Thus in phase 2 epidermis, the higher the mitotic rate the more mitotic-cycle cells there are per unit area of the skin surface, and according to the N ratio, the more post-mitotic cells there must also be, even though these cells are ageing and dying more quickly. It follows that, in phase 2, the higher the mitotic rate the thicker the epidermis, an extreme being reached, for instance, in an epidermal papilloma, beneath which the dermo-epidermal junction is particularly deeply folded.

However, the hyperplastic state, like the normal state, is fully stable; cell gain continues to be exactly balanced by cell loss. The increase in epidermal cell number and in epidermal thickness takes place only during the short period while the epidermis is passing from the one stable state to the other. Thus, for instance, if human epidermis becomes hyperplastic, as at the onset of psoriasis, and the raised mitotic rate is offset by a reduction in the post-mitotic life span from some 2 or 3 weeks to only about 4 days, then only during the first 4 days will the number of epidermal cells increase and the epidermis thicken. After 4 days the rate of cell ageing and death comes into balance with the rate of cell gain, and a stable plateau is once more reached.

It is important to note that, when the mitotic rate increases in this way, not only do the newly-forming post-mitotic cells age and die more quickly, but so also do the already existing post-mitotic cells, which up to that time had been ageing slowly. The post-mitotic ageing rate, like the mitotic rate, responds immediately to changes in the cellular environment.

When hyperplastic epidermis reverts to normal the reverse process occurs. As the mitotic rate falls, the number of basal mitotic-cycle cells decreases, and the rate of post-mitotic ageing and keratinization slows. The mass of extra post-mitotic cells which had been built up and which had been ageing and dying in about 4 days, now need the normal 2 or 3 weeks to do so (Argyris, 1977; Bullough and Stolze, 1981, unpublished data), and consequently the time taken by the epidermis to return to its normal thickness is much longer than is required for the increase in thickness at the onset of the hyperplasia.

3.1.4 *Epidermal wound healing*

The source of the extra epidermal cells needed to make good the loss caused by wounding must now be considered. In the type of hyperplasia considered above the basal epidermal cells were only damaged but in a typical wound they are lost altogether and the underlying dermis is exposed.

In such a case the re-establishment of the basal cell layer is achieved by cell migration (see 4.4.2). Cells from the wound periphery move out across the wound cavity until it is again completely covered by epidermis. These cells do not travel across the surface of the exposed dermis, where, being unprotected by an outer keratin layer, they would die of desiccation. They move beneath the surface at a level where both water and nutrients are available, cutting by enzyme action through the obstructing collagen fibres as they go (Winter, 1972).

The extra cells needed for this migration are produced by the increased mitotic activity of the basal epidermal cells at the wound edges in response to the weakened chalone action in that region. As they pass through the dermis they, of course, remain in dermal contact, with the result that, as they settle into their places, they undergo active mitosis. With little or no chalone inhibition their mitotic rate is particularly high, and while some of the daughter cells move out still further into the wound gap, others are pushed distally to become post-mitotic, to keratinize and to die. As usual with a high mitotic rate these distally moving cells age rapidly, but in each region, within the limited time available before cell loss again equals gain, the high mitotic rate results in the production of a mass of new cells. The new epidermis so produced is, of course, hyperplastic.

The healing of a large wound may take many days, the new epidermis being produced zone by zone as the migrating cells progress across the gap. Behind the converging wound edges the mitotic rate falls towards normal as the chalone effectiveness increases (Bullough and Laurence, 1960a), and simultaneously the post-mitotic cell life span lengthens. The epidermis then remains thickened for some 2 or 3 weeks until the extra numbers of supra-basal cells that were produced are finally cornified and a normal epidermal structure is again attained.

The healing of a small wound, such as a pin prick, may take place entirely by cell migration. It is known that the epidermal chalones exert their influence over a distance of some 0.5–1.0 mm, and it seems that in a wound of not more than this width the chalones continue to dominate all the cells, so that there is no increase in the mitotic rate (Bullough and Laurence, 1960a).

3.1.5 G_1 *chalone action in hyperplasia*

It is still not certain whether, after extensive epidermal damage, the rise in the mitotic rate that signals the beginning of hyperplasia is caused by the local fall in the G_1 chalone concentration or by the reduced ability of the damaged cells to respond adequately to the chalone that is there. It is probably a combination of the two.

Early experiments indicated that in the neighbourhood of a wound the concentration of the epidermal G_2 chalone may fall to about half normal

(Bullough, 1969); later experiments have shown that after epidermal damage the amount of G_1 chalone that can be extracted falls almost to zero (Elgjo and Devik, 1978).

More recently Marks *et al* (1978) have concluded that, at least during the first 2 days of increased mitotic activity after epidermal damage, the affected cells lose their ability to respond to the G_1 chalone. The cellular damage is such that it 'switches off the chalone sensitivity', although it must be noted that it does not switch off the cell response to the pro-mitotic message of the mesenchymal factor. After 2 days the epidermal cells, still with a raised mitotic rate, recover their ability to respond to the G_1 chalone.

Regarding the epidermal G_2 chalone, a similar situation may exist. It is on the second day after wounding that an increase in the stress hormone concentration greatly increases the competence of the damaged epidermal cells to respond to the G_2 chalone (Bullough, 1969).

These observations fit well with recent descriptions of the responses of guinea pig epidermis to chronic irritation (Bullough and Stolze, 1981, un-published data). The immediate response, lasting for about 2 days, is a massive increase in the mitotic rate to more than $\times 20$ normal. After day 3, when the epidermal thickness has doubled, the mitotic rate falls to only about $\times 6$ normal at which level it remains for as long as the irritation is maintained. The initial greater response may be caused both by the fall in the G_1 chalone concentration and the failure of the damaged cells to react to whatever chalone remains, while the later lesser response may be due mainly to the reduced G_1 chalone concentration to which the cells have recovered their ability to respond. Other factors possibly influencing this later response could be the better protection from irritation provided by the thicker epidermis and the greater quantity of chalone produced by the greater number of epidermal cells.

This two-phase response to damage has also been described for other tissues. In damaged dermis the initial increase in the mitotic rate, lasting in this case for about 4 days, is about $\times 150$ normal, while the later lesser response, which is maintained for as long as the damage continues, is only about $\times 50$ normal (Bullough and Stolze, 1981, unpublished data).

3.2 The effects of stress

In conditions of chronic stress the mass of the hyperactive adrenal glands increases greatly, and the increased output of the two adrenal hormones powerfully strengthens the actions of the two epidermal chalones. Clearly the adrenal hormones do not contribute to the basic function of the cellular homeostatic mechanism, that is they do not influence the balance between cell gain and cell loss. Their action is to reduce equally the rates at which cells enter and leave the epidermal system. The question now arises as to the possible biological significance of this action.

3.2.1 *Stress and cell production*

Although any stressful situation, however caused, reduces the rate of cell flow through the epidermis, in a wild animal the most important single cause of chronic stress is the annual cold (or hot) season which is also accompanied by food shortage. It is well established that chronic hunger causes a considerable increase in the size and secretory activity of the adrenal glands, and that this in turn leads to a great reduction in the rate of replacement of the epidermal and other types of tissue cells (Bullough and Eisa, 1950).

Rytömaa (1976b) has calculated that in a normal man, made up of approximately 10^{14} cells, the amount of tissue lost and replaced per day is about 1–2% of his total body weight, so that in 50–100 days a man produces a mass of new tissue equal to his own mass. Even though many of the materials of the old dead cells may be reclaimed and re-used, this represents a heavy daily metabolic demand, which in conditions of starvation may be difficult, or even impossible, to meet. It could be a matter of life or death that the rate of tissue loss and tissue replacement be reduced, and it is known that, within limits, such a reduction has no adverse effects.

The most important experiments on the effects of hunger on cell replacement have been those done on rats and mice. In these animals a restriction of food intake to two-thirds of normal induces chronic stress, causes an increase in the adrenal mass (Tannenbaum and Silverstone, 1957), and reduces the rate of cell flow through the epidermis and the sebaceous glands to a quarter or fifth of normal (Bullough and Ebling, 1952). In these circumstances the epidermis becomes hypoplastic but functions normally, and the evidence indicates that a similar situation develops in the other body tissues. Cells that would normally have aged and died continue to function properly for remarkably long periods of time, and indeed when animals are kept throughout the whole of their lives on a restricted diet with constant stress they actually remain in better health and live much longer than do their fully-fed littermates (7.3.1).

Regarding human epidermis, one result of such chronic stress was shown in the Japanese prisoner-of-war camps when prisoners who normally suffered from psoriasis never developed this hyperplastic condition until after their release. Their epidermis was evidently unable to escape from its chronic hypoplasia.

It can therefore be suggested that the main function of the adrenal hormones in relation to the chalone control mechanisms may be to provide a regulatory link between the external conditions, especially the amount of food available, and the amount of new tissue that is created daily.

A further important point relates to wound healing. One obvious result of wounding may be stress, which would then be expected to slow the healing process by limiting the rate of new cell production. However, with reduced chalone action in the epidermal cells lateral to the wound, there is also a reduced stress hormone influence. During the first 2 days when mitotic activity is maximal, stress hormone action is minimal, even at artificially high hormone concentrations. Thereafter, as the cells regain their responsiveness to chalone action, but while the chalone concentration is still subnormal, the mitotic inhibition caused by the stress hormones is proportional to the chalone concentration. Thus, even though the stress may cause some delay, it does not usually seriously inhibit the process of wound healing.

3.2.2 *Age and stress*

Further evidence of the effects of stress on the rates of epidermal cell gain and cell loss comes from studies of ageing animals (Bullough, 1949). The life of a mouse, like that of other mammals, passes through at least four main phases: the immature age (1–2 months) when the animal is still growing; the mature age (3–12 months) when the body weight is normal and the animal is sexually active; the middle age (13–c. 20 months) when the body weight usually increases considerably; and the senile age (> c. 20 months) when the animal becomes emaciated and feeble.

The first age, when the mitotic rate is relatively high, and the last, when it is low, require no particular comment, but the increase in the mitotic rate between the mature age and the middle age, which in the male mouse may be abrupt, is remarkable. This change is accompanied by an equally abrupt change from an active mode of life to one that is much quieter and lazier. Quite suddenly the daily physical activity may be reduced by almost half, and this results in heavy fat deposition. Evidently the middle-aged animals lead much less stressful lives with the result that, both in the epidermis and in the other tissues examined, the mitotic rate (and therefore the rate of cell ageing and loss) is greatly increased. There could be some connection between this and the fact that this is the age at which cancer commonly begins.

3.3   A chalone negative feedback mechanism

The epidermis is a massive tissue, and consequently it must produce relatively large amounts of its two chalones, which, being remarkably stable, may then escape into the circulation. In fact the G_2 chalone has been extracted from the dermis (Elgjo, 1976) and even more significantly from human urine (Bullough and Laurence, 1971, unpublished data). The question, therefore, arises whether these chalones, distributed via the bloodstream, can produce a significant effect, one region of the epidermis interacting with another.

The evidence is clear that this does not normally happen to any significant extent, and that the epidermal chalone control mechanism is strictly local in its operation. Areas of high mitotic activity, such as the mouse footpads, retain their sharp boundaries with adjacent areas of low mitotic activity (Bullough and Stolze, 1981), and even when the skin is wounded the stimulus to a higher epidermal mitotic rate does not extend further than 1-2 mm from the wound edge (Bullough and Laurence, 1960a; Winter, 1972).

However, it is theoretically possible that after very extensive skin damage or loss there could be such a fall in the systemic chalone concentration as to cause a rise in the mitotic rate of the whole of the undamaged epidermis. In such circumstances a simple negative feedback effect could develop.

3.3.1   *General chalone loss*

The fact that a massive loss of tissue can have this effect is shown after partial hepatectomy, which causes not only a general rise in the mitotic rate of the whole of the remaining liver, but also a similar rise in a liver fragment implanted at a distance or in the untouched liver of a parabiotic twin (6.2.3); after the removal of one kidney, which leads to the growth of the other; and after massive haemorrhage, which leads to extra erythrocyte and granulocyte production (8.2.1). These responses evidently depend on the abnormal development of simple negative feedback mechanisms. When the tissue mass becomes inadequate in relation to the total body space, the systemic chalone concentration falls to such a low level that there is an increased rate of chalone loss from the surviving tissue remnant, which consequently has a reduced chalone concentration. The subsequent increase in the mitotic rate is then proportional to the amount of tissue lost (6.2.3).

Presumably this type of response can only develop in tissues that are massive enough to produce a significant systemic chalone concentration, and it is probable that the epidermis is such a tissue. However, any attempt to demonstrate this by destroying a large enough area of epidermis would obviously be fatal, and the experiment is therefore impossible.

3.3.2 *General chalone excess*

The opposite experiment, to determine the effect of a general chalone excess, is however easily possible. The simplest demonstration of the consequences of increasing the systemic concentration of the epidermal chalones is when, after a subcutaneous chalone injection, there follows a general mitotic depression over the whole epidermis.

More interesting, however, is the effect of an excessive number of epidermal cells, a situation that develops during the growth of an epidermal tumour. This results in a steady increase in the systemic concentration of the epidermal chalones and consequently a steady decrease in the rate of epidermal cell production over the whole body surface (9.2.4). Ultimately the whole epidermis is reduced to the phase 1 hypoplastic condition.

3.4 Summary and conclusions

The stratified structure of the epidermis is evidently maintained by the basal pro-mitotic influence of the dermal mesenchymal factor and the distal anti-mitotic and pro-keratinization influence of the G_1 chalone. A basal cell responds to its environment by mitosis; a supra-basal cell responds to its environment by keratinization and ageing. Cellular stratification is also ensured by the fact that epidermis is a non-vascular tissue. The lack of intra-epidermal blood vessels means the absence of intra-epidermal connective tissue, which would stimulate the production of nests of supra-basal mitotic cycle cells and so complicate the pattern of cellular maturation.

3.4.1 *The chalone control mechanism*

Epidermal cellular homeostasis is ensured by a complex chalone control mechanism and not by the simple negative feedback mechanism that was originally proposed. The chalone mechanism determines the mitotic rate, the process of post-mitotic cell maturation (keratinization), and the rate of post-mitotic cell ageing; it responds to the actions of at least three primary messenger molecules, the mesenchymal factor and the two tissue-specific chalones; it includes those still unknown intracellular secondary mechanisms that transmit the trigger actions of the primary messenger molecules to the nucleus and the genes; and it is connected to the adrenal hormone system, probably via the intracellular secondary mechanisms.

Only in such massive tissues as the liver, and in the pathological condition that follows excessive tissue loss, does a simple chalone negative feedback mechanism become apparent, based on changes in the systemic chalone concentration.

3.4.2 *The rate ratio*

Cell gain always equals cell loss because of the constancy of the ratio between the mitotic rate of the basal epidermal cells and the ageing rate of the supra-basal keratinizing cells. This is the R ratio. It is suggested that, with the concentration of the mesenchymal factor being constant in any one area of skin, the rate of new cell production is determined by the basal concentration of the G_1 chalone, while the rate of post-mitotic cell ageing is determined by the supra-basal concentration of the G_2 chalone. The value of the constant R ratio may thus be determined by the value of the ratio between the G_1 and G_2 chalone concentrations. If this is true it would explain why there are two epidermal chalones and not just one.

In any one mammalian species the value of the R ratio may be constant over the whole epidermis and may thus determine the general epidermal thickness: the slower the rate of post-mitotic cell ageing relative to the mitotic rate (that is the higher the G_2 chalone concentration relative to the G_1 chalone concentration), the thicker will be the epidermis. Thus the value of the R ratio may be species-specific.

However, in any one mammalian species the epidermal thickness is not constant but differs markedly from body region to body region. These differences are not caused by any change in the R ratio, but are due to one or more of a number of dermal influences (4.2.2, 4.5.1). Of these the first is the local concentration of the mesenchymal factor. Experiments on epidermal transplantation have shown that all the epidermal cells are equipotential (which confirms that in all body regions the value of the R ratio is the same), and that a higher (or lower) local mitotic rate, with a locally thicker (or thinner) epidermis, is due to the higher (or lower) local concentration of the mesenchymal factor.

3.4.3   *Post-mitotic cell ageing*

The process of supra-basal cell ageing remains a mystery, and its manner of control can only be guessed. It is probable that it is related to the equally mysterious process of the ageing of the whole animal (see 7.3.1), and it remains an important field for future study. It is, however, already clear that the process is closely controlled, that it is tissue-specific in its timing, and that it is directly linked to the mitotic rate. The controlling mechanism, whatever its nature, is an integral part of the epidermal chalone mechanism.

The possibility has been discussed that the G_2 chalone may determine by inhibition the rate of cell ageing, and if this proves not to be the case, then it may be necessary to search for some other tissue-specific messenger molecule that does so. For the moment it may be proposed that the high basal G_2 chalone concentration may prevent the ageing of the basal cells, while the reduced supra-basal concentration may permit and control the ageing of the supra-basal cells.

If this is true then the nature of the link between the mitotic rate and the post-mitotic ageing rate would be simple: both the G_1 and G_2 chalone actions are strengthened by the adrenal hormones during stress and weakened after any form of tissue damage, so that the mitotic rate and the post-mitotic ageing rate would be equally affected. At the moment, however, the evidence for this suggestion is indirect.

Chapter 4

Epidermal cellular organization

4.1	*Epidermal thickness*		55
	4.1.1	*Basal cell crowding*	55
	4.1.2	*Dermo-epidermal adhesion*	56
4.2	*Basal cell organization*		57
	4.2.1	*Basal cell extrusion*	57
	4.2.2	*Lateral tension in the basal layer*	58
	4.2.3	*Hyperplasia and lateral tension*	59
	4.2.4	*The dermo-epidermal junction*	60
	4.2.5	*Vertical mitosis*	61
	4.2.6	*Computer predictions of baseline tension*	62
	4.2.7	*Responses to skin stretching*	62
4.3	*Supra-basal cell organization*		63
	4.3.1	*The columnar cell pattern*	64
	4.3.2	*The roof-tile cell pattern*	66
	4.3.3	*Cellular disorder*	66
	4.3.4	*The stratum corneum*	67
4.4	*Cellular re-organization after wounding*		67
	4.4.1	*The basal cell layer*	68
	4.4.2	*The supra-basal cell layers*	69
4.5	*Summary and conclusions*		70
	4.5.1	*Local epidermal thickness*	70
	4.5.2	*Intra-epidermal tension*	70
	4.5.3	*The fate of the basal cells*	71
	4.5.4	*Supra-basal cell movement*	72

Although the epidermis may seem to be one of the simplest tissues, with the mitotic cells basal and the keratinizing cells distal, it has in fact a highly organized structure, or cellular architecture, that is created by the precise way in which the cells move and change shape as they mature and rise towards the surface.

From this dynamic point of view there are three main epidermal regions: the basal sheet of mainly (phase 1) or wholly (phase 2) mitotic-cycle cells that is firmly attached to the dermal surface; an intermediate zone in which the newly-formed post-mitotic cells, although loosely attached to each other, are free enough individually to be able to move sideways as well as distally as they manoeuvre into their proper final positions; and an outer zone in which the flattened keratinizing cells are tightly glued together and to the distal squames.

Thus there is a fixed *stratum basale*, an intermediate *stratum mobile*, and an outer *stratum compactum*, and in the first and last of these layers the cells are organized into particular patterns.

In the various types of mammalian skin the epidermal structure varies in only two main particulars: it may be thicker or thinner, and its inner and outer cells may be arranged in one pattern or another. These differences are interrelated.

4.1 Epidermal thickness

The thickness of the epidermis is determined by three main factors, of which two have already been described above. The first and most basic factor is the value of R and N ratio constants (3.1.1), which may vary between species. The second is the mitotic rate (in phase 2 epidermis), which may vary from region to region of the body according to the strength of action of the pro-mitotic mesenchymal factor in the dermis (3.1.3). And the third, now to be discussed, is the degree of crowding of the cells of the stratum basale.

4.1.1 *Basal cell crowding*

The number of basal epidermal cells per unit area of skin varies greatly between different regions of the body and between different mammalian species. The more crowded the cells are per unit area of baseline the more columnar they are in form, and in addition, the baseline itself may be deeply folded to accommodate even greater cell numbers.

As one example there are about five times as many basal epidermal cells per unit area of mouse foot pad as there are per unit area of mouse ear (Bullough and Stolze, 1981, unpublished data). Then, according to the constant N ratio (3.1.1), the more basal cells there are, the more supra-basal cells there must also be, and the thicker the epidermis. Thus the mouse foot pad epidermis is much thicker than is the ear epidermis (see 4.2.2).

Studies of the situation in the mouse have shown that the main reason for any local difference in basal cell crowding, and therefore also in cell shape, is a local difference in the strength of the basal cell grip on the dermal surface (Bullough and Stolze, 1981, unpublished data). The stronger the basal cell grip the more the cells are able to resist detachment and extrusion distally when new basal cells are produced by mitosis. The rise in tension within the basal cell layer, created by mitotic activity, results from the fact that each mitosis usually occurs with its axis parallel to the epidermal baseline, so that both the new daughter cells tend to remain in the basal layer (but see 4.2.5). It is some closely adjacent basal cell, which is either in the early A_1 phase (phase 1 epidermis) or in the intermitotic G_{1b} phase, that is usually forced out to create the necessary space. The greater the grip of the basal cells, the higher the tension needed to displace these cell types, the more crowded and columnar the basal epithelium becomes, and the thicker the epidermis.

4.1.2 *Dermo-epidermal adhesion*

The adhesive force that binds the epidermis to the dermis has been studied, mainly in human epidermis, by measuring the suction pressure required to raise an epidermal blister (Kiistala, 1972; Leun *et al*, 1974). This has led to the conclusion that, over a wide range of suction pressures, $pt = A$ (where p is the pressure, t is the time taken for the blister to form, and A is the strength of the dermo-epidermal adhesion), and that the value of A decreases exponentially with increasing temperature. These relationships are such as to suggest that the epidermis is held to the dermis by the viscosity of a fluid interface, as two sheets of glass are held together by a water interface.

In such a physical system the force required to break the bond is time-dependent; a strong force acts after only a short time while a weak force acts equally successfully but after a long time. However, it has been found that at very low suction pressures the basal epidermal cells are able to resist detachment indefinitely (Leun *et al*, 1974). This has been taken to indicate the existence of other weaker adhesive factors, in particular a system of fibrous connections embedded in the glycoprotein ground substance between the bases of the epidermal cells and the superficial dermis (Omar and Krebs, 1975), and a process of re-adherence by which the basal cells are able to repair continuously the damage done by the suction to the dermo-epidermal junction.

In addition, the strength of attachment of the basal cells must depend on the area of contact between their bases and the dermal surface. In some body regions, for instance in pig (Winter, 1972) and man (Schmidt *et al*, 1974), each basal cell possesses large numbers of tiny finger-like projections, called '*Füsschen*', which are interdigitated into the dermal surface. In the human palm of hand and sole of foot it is these that help to resist any shearing forces.

4.2 Basal cell organization

Important as these conclusions are to an understanding of dermo-epidermal adhesion, the forced detachment of a single cell by the tension generated by an adjacent mitosis depends on more complex considerations. In particular, since the two daughter cells formed by each mitosis usually retain their basal grip while an adjacent intermitotic G_{1b} cell is displaced, it is clear that the strength of the basal cell grip must vary with the phase of the mitotic cycle.

4.2.1 *Basal cell extrusion*

Evidence from mouse epidermis (Bullough and Mitrani, 1976) shows that S phase cells (and, therefore, also G_{1c} cells) are always basal, as also are cells in mitosis (and therefore also in G_2) and in G_{1a}. Thus throughout the whole mitotic process a cell is not forced from its basal attachment. The only cells that are forced out are, in phase 1, those in the early A_1 phase that contain some keratin precursors, and in both phase 1 and phase 2, those that are in the intermitotic G_{1b} phase.

The only exception to this rule is seen in those mitoses which in certain circumstances may become angled vertically so each distal daughter cell is directly extruded without ever being in contact with the dermal surface (Bullough and Mitrani, 1978). The reasons for this are discussed in 4.2.5.

The conclusion is that some feature of the cell membrane that is involved in basal cell adhesion must vary with the cell cycle, and since the basal cells are also linked sideways to each other through the desmosome connections, the strength of this lateral linkage must also be taken into consideration.

4.2.2 *Lateral tension in the basal layer*

The pressure that acts on the basal epidermal cells comes from two main directions: the vertical direction caused by the elastic back pressure of the stratum compactum and the corneum acting against the resistance of the dermis, and the lateral direction caused by the degree of cell crowding in the basal layer. If the pressures coming from these two directions are roughly equal the result is a cubical epithelium, but if, as is often the case, the lateral pressure is the greater the result is a columnar epithelium.

The lateral tension within the basal cell layer has two components: there is the constant tension which derives from the normal degree of cell crowding, and which is itself a function of the grip of the cells at its weakest point in the intermitotic G_{1b} phase; and there is the additional local and episodic tension which results from a mitosis and which, after a short interval, is relieved by the extrusion of a local G_{1b} phase cell. In other words, the lateral tension in the basal cell layer builds up by mitotic activity to a steady level, beyond which any further increase in tension due to further mitotic activity is adequate to force the intermitotic cells to release their basal grip.

Estimates have been made of the degree by which lateral tension exceeds vertical tension both by measuring the degree of lateral compression of the basal cells (the height:width ratio of the nuclei) and by counting the number of basal cells per unit area of baseline (Bullough and Deol, 1975; Bullough and Stolze, 1981, unpublished data). Considerable differences have been found. Thus in mouse ear epidermis the nuclear height : width ratio is 1.0, which means that the pressure is roughly equal in both directions, and that the intermitotic G_{1b} phase cells are easily detachable; the number of basal cells per $0.01\,mm^2$ baseline area is about 130. In contrast, in mouse sole-of-foot the nuclear ratio is almost 4.0, which means that there is a constant high lateral pressure due to the tighter baseline grip of the G_{1b} phase cells; the number of basal cells per $0.01\,mm^2$ baseline area is about 325. According to the N ratio, the more basal cells there are the more distal cells there must also be, and the result is the thicker sole-of-foot epidermis.

To this is added a further thickening effect: in the sole-of-foot the dermo-epidermal junction, instead of being flat as it is in the ear, is deeply folded. Consequently, by reference to a unit area of skin surface instead of to a unit area of dermal surface, the basal cell number of 325 is doubled to about 650. Thus in sole-of-foot the basal cell number per unit skin area is almost five times that in the ear, which is reflected in the fact that sole-of-foot epidermis is about seven times as thick as is ear epidermis. The fact that the increase in thickness is greater than might be expected is due to the high columnar shape of the basal and supra-basal cells and to the thicker stratum corneum.

In human body epidermis, as also in guinea pig, there is an inter-mediate situation (Bullough and Stolze, 1981, unpublished data). The nuclear height : width ratio is about 1.7, and the basal cell number per 0.01 mm^2 of baseline is about 160.

These differences are quite independent of the mitotic rate, but like the mitotic rate they are dermis-dependent. The classical transplantation experiments of Billingham and Silvers (1968) showed that epidermis taken from one body region, such as the ear, and implanted on the dermis of another body region, such as the sole-of-foot, always acquires the epidermal characteristics of its new site, but that if some of the original dermal bed is transplanted with the epidermis its characteristics remain unchanged. Thus the grip of the basal epidermal cells, their degree of crowding, and the consequent thickness of the epidermis are determined by the dermis.

4.2.3 *Hyperplasia and lateral tension*

As already explained, when the epidermal mitotic rate rises not only does each mitotic-cycle cell enter mitosis more often, but the number of cells involved in the mitotic cycle also increases. These extra cells are accommodated in one of two different ways. In the first, which is typical of mouse epidermis, they form one or more supra-basal cell layers; because of the reduced effectiveness of the G1 chalone, mitotic activity is no longer confined to the basal cell layer. In the second, which is typical of human epidermis, the extra mitotic-cycle cells are accommodated on an increased baseline area created by the deeper folding of the dermo-epidermal junction, although here too, as in the mouse, some of these extra cells are supra-basal.

However, in a wide range of epidermal types it has been found that an increased mitotic rate does not lead to any increase in the number of cells per unit area of baseline (Bullough and Stolze, 1981, unpublished data), and thus it is clear that in hyperplasia the baseline grip on the epidermal cells does not change and there is no increase in lateral tension in the basal cell layer. This may be of critical importance because if the raised mitotic activity in hyperplasia did lead to increased lateral tension there might be a breakdown in epidermal structure, including perhaps an invasion of the dermis such as occurs in an epidermal carcinoma.

In fact any such rise in lateral tension in the basal cell layer is prevented by three distinct mechanisms. The first has already been emphasized: as new cells are formed faster by mitosis, neighbouring intermitotic cells are extruded faster into the distal cell layers. The other two mechanisms are more subtle: one is the deeper folding of the dermo-epidermal junction typical of human epidermis (4.2.4), and the other is the increasing incidence of basal mitoses that are angled vertically instead of horizontally (4.2.5; 4.2.6).

4.2.4   *The dermo-epidermal junction*

The reason for the increased folding of the dermo-epidermal junction in hyperplasia is not clear. The first impression is that the greater number of basal mitotic-cycle cells that are formed may create an extra degree of lateral tension, which is then the driving force for the deeper penetration of the epidermal folds into the dermis. On this view, the system would remain under extra tension until the epidermal mitotic rate returned to normal, when the compressed dermis would force the dermo-epidermal junction to unfold. That this cannot be true is shown by the absence of any increased lateral tension in the basal cell layer of hyperplastic epidermis.

An alternative explanation is that the folding is initiated by the dermis. After any form of skin damage the dermis also thickens, first, due to oedema and to the wide distension of the superficial capillary loops, and second, due to dermal hyperplasia, which affects all cell types including those of the capillaries. In the normal epidermis of guinea-pig or man the capillary loops lie within dermal ridges which protrude into the base of the epidermis. In hyperplasia it is these capillary loops and their surrounding dermis that show the greatest degree of growth, so that it is the dermal ridges as they increase in size that seem to force themselves deeper into the base of the thicker epidermis. Simultaneously, the capillary loops merge together in groups by the elimination of the intervening epidermal tissue; thus the dermal ridges become fewer in number as they become wider and deeper.

This lends support to the theory that it is some change in the superficial dermis that is a basic cause of psoriasis, which is also characterized by a deeply folded dermo-epidermal junction (Braun-Falco and Christophers, 1974).

Whatever may be the ultimate cause of the deeper baseline folding with increasing mitotic activity, its value is clear. It provides a greatly increased area on which the extra numbers of mitotic-cycle cells can be based without any disruptive increase in the lateral tension within the epidermis. However, this is a safety mechanism which, valuable as it may be, is not essential to epidermal stability; in mouse ear epidermis the baseline does not fold and the extra mitotic-cycle cells merely pile on top of each other to form two or more layers.

4.2.5 *Vertical mitosis*

The other mechanism that preserves epidermal stability in hyperplasia is more important. As the mitotic rate increases the intermitotic G_{1b} phase shortens, and consequently the higher the mitotic rate the less probable it becomes that a G_{1b} phase cell will be locally available to be extruded when a basal cell divides. It has been calculated (Bullough and Mitrani, 1976) that in the basal layer of normal mouse epidermis the ratio G_{1b} cells : mitotic cells is roughly $10:1$, and that in hyperplasia this may change to $1:1$. Even with a ratio of $10:1$ many mitoses will occur where no extrusible G_{1b} cell exists; in hyperplasia this problem becomes acute.

The pressure created by a dividing cell does not build up gradually; there is a sudden increase in cell volume at the beginning of the mitosis, or M phase, probably due to rapid water intake. This is the moment when the mitotic axis is determined: in the presence of an extrusible G_{1b} cell the axis is horizontal to the baseline, and both daughter cells remain basal; in the absence of a G_{1b} cell the axis is vertical to the baseline, so that the distal daughter cell has no basal attachment and is directly extruded (Bullough and Mitrani, 1978). It is a mystery how, within the few minutes required for the expansion of the dividing cell, the presence of an adjacent G_{1b} cell can be 'sensed'; certainly the subsequent extrusion of this cell must take several hours. It is possible that a G_{1b} cell is relatively flaccid, and that in its absence the surrounding cells are so turgid that the mitotic cell can only expand outwards. In other words, the angle at which a basal cell divides depends on whether the potentially available space is horizontal or vertical.

Observation shows that in the basal cell layer of normal mouse epidermis with a low mitotic rate the vertical mitoses average about 10%, while in hyperplasia they increase to about 25% (Bullough and Mitrani, 1978); in normal guinea-pig epidermis with a higher mitotic rate they average about 40%, while in hyperplasia they increase to about 60%; for normal human epidermis data are lacking but in psoriatic hyperplasia, a figure of about 50% has been found (Bullough and Stolze, 1981, unpublished data).

These relatively high percentages of vertical mitoses seen in guinea-pig and human epidermis require comment. In contrast to the mouse, which has a cubical basal epithelium, these two species have a columnar basal epithelium, and if the axis of a mitosis is determined by the shape of the space available to it, it must follow that more vertical mitoses will be present in a columnar than in a cubical epithelium. In fact it has been found that the effects on the incidence of vertical mitosis of a high mitotic rate and of a columnar epithelium are additive (Bullough and Stolze, 1981, unpublished data).

4.2.6   *Computer predictions of baseline tension*

The relations between the changing mitotic rate, the incidence of vertical mitosis, and the potential pressure in the basal epidermal cell layer have been analysed by computer (Bullough and Mitrani, 1978). It has been found that, over a wide range of real values, the potential lateral pressure caused by the lack of extrusible G_{1b} cells remains low, but that when the G_{1b} half-life shrinks into the 15–10 hour range, the pressure rises rapidly and the whole system becomes potentially unstable. It is precisely within this range that observation shows there is a sudden rapid rise in the proportion of vertical mitoses.

The computer analysis also predicted that if the duration of the whole mitotic cycle were to fall below about 20 hours, the portion of the cycle spent in the G_{1b} phase would become so short that the chance that an extrusible cell would be available alongside a mitosis would be practically zero. In this situation all the mitoses would have to take place in the vertical axis. In epidermal hyperplasia this point never seems to be reached, but in an epidermal derivative, the hair bulb, the mitotic rate is so extremely high that the cell cycle duration is only about 12 hours. In these cells all the mitoses are indeed angled vertically.

4.2.7   *Responses to skin stretching*

This question of a potential rise in baseline tension following an increase in the mitotic rate raises the contrary question of a potential fall in tension following the forced expansion of the baseline. This situation may occur naturally and gradually, for instance during pregnancy, or it may be studied experimentally when, however, the stretching of the skin is immediately maximal (Squier, 1980). In such an experimental situation the sudden tension causes epidermal damage by disturbing the inter-cellular connections and distorting the intercellular spaces, and the result is an increased mitotic rate followed by hyperplasia. This experiment is clearly unnatural since the tension is maintained and the system never comes to rest.

The expansion of the skin in pregnancy provides a better answer to the question since it is gradual and no epidermal damage ensues. The slow increase in area of the epidermal baseline is simply accompanied by an equivalent increase in the total number of basal epidermal cells, so that the normal basal cell number per unit area of baseline is maintained. This occurs because, with a lower than normal lateral tension in the basal cell layer, the intermitotic G_{1b} cells are not always forced out when a mitosis occurs. Enough of these cells are retained on the baseline to maintain the normal lateral tension and thus the normal shape of the basal cells.

After pregnancy, during the shrinkage of the skin, this process is reversed with a greater than normal number of G_{1b} cells being extruded as the baseline area is reduced.

4.3 Supra-basal cell organization

The supra-basal cells of the epidermis are arranged in two main layers: the inner stratum mobile in which the cells move relative to each other, and the outer stratum compactum in which the cells and squames are tightly glued together. Cells entering the stratum mobile are roughly rounded, indicating equal pressure from all sides, and are attached to each other by desmosomes, which give the cells their well-known spinous appearance. These desmosomal connections can evidently be easily broken and easily re-established as the cells move outwards and side-ways, and as they enlarge and flatten while manoeuvring into their appropriate places within the epidermal cell pattern. The movements of the stratum mobile cells have never been directly observed but the precision and orderliness of their migrations can be deduced from the resulting cellular architecture of the stratum compactum.

When approaching their final positions the cells of the stratum mobile flatten and expand to such a degree that each comes to cover some 10–15 of the basal cells. This plate-like expansion involves a considerable increase in volume and in dry weight. According to MacKenzie (1972, 1980) this flattening is an 'active property of the individual cells and is not the result of cell population pressure'. It is also seen in dissociated cells *in vitro*, in chaotic epidermal carcinomata, and in embryos in which no stratum corneum is present to exert a back pressure.

The flattening process is completed as the cells come to lie in their final positions, in which they sometimes form stacks or columns, sometimes overlap each other like the tiles of a roof, and sometimes are crowded together in apparent complete disorder. There is no indication that any one of these cell arrangements produces a more efficient type of stratum corneum. However, it is the thinnest type of corneum that is structured in cell columns and the thickest that shows cellular disorder, and it is therefore possible that the columnar arrangement is more efficient in enabling a thinner corneum to function as well as a thicker corneum.

4.3.1 *The columnar cell pattern*

This type of cell arrangement in the stratum compactum is typical of a relatively thin epidermis with a low mitotic rate. It is probably always present in phase 1 epidermis, and it is also seen at the lower end of phase 2. The stratum mobile is also thin but each cell remains in it for a relatively long time (2-3 days in mouse ear epidermis) as it flattens and moves into a central position beneath one of the overlying cell and squame columns. Viewed from above the flattened cells are roughly six-sided so that the columns of cells fit together in a honeycomb pattern (MacKenzie, 1969, 1972; Christophers, 1971).

Detailed studies of the columns show that the edges of the cells and of the squames in neighbouring stacks overlap each other alternately; as Fitzgerald (1977) has described it, 'the squames from adjacent stacks are perfectly riffle-shuffled, their edges alternating with mathematical precision'. In two-dimensional sections this arrangement seems relatively simple, but in three-dimensional reality it is complex, each stacked column having not two but five or six boundaries. It obviously requires a subtle pattern of new cell accretion to ensure that each and every column shall possess the observed boundary arrangements. Menton (1976) has stressed the close similarity between this method of cell packing and that of soap bubbles, which, however, do not produce the vertical column alignment.

The method of cell addition to the base of the stratum compactum has been described by MacKenzie (1972): 'A cell which has flattened into line beneath a particular column lies below cells previously added to the adjacent columns, and therefore interdigitates beneath them forming a step like boundary at its lateral margins. When the next cell which is to flatten moves up from below, the presence of this step may tend to guide it away from the territory of the first column. The cell therefore flattens into line beneath an adjacent column.' In other words, the configuration of the underside of the stratum compactum may act as a template to guide the movements of the new cells coming from below, each new cell entering the pattern at a point where an unfilled cavity exists.

This columnar pattern of cell stacking is maintained only so long as the rate of basal cell mitosis is such as to add not more than one new cell per day to each column (Christophers et al, 1974). Any rise above this figure seems to result in too many cells of the same age and maturity competing for the same position, with the result that the pattern breaks down. This cellular competition is aggravated by the fact that the higher the mitotic rate, the faster the cells move through the stratum mobile, and the shorter is the time available for them to manoeuvre into position. If the increase in the mitotic rate is due to damage, yet another complication arises from the fact that the cells become swollen so that the stratum mobile cells become too large to fit beneath the existing columns (Bullough and Stolze, 1981, unpublished data).

It is interesting that in the underlying basal cell layer the mitoses tend to occur beneath the edges of the overlying cell columns, and also that the intermitotic cells that are forced out into the stratum mobile tend to come from this same peripheral region (MacKenzie, 1970; Christophers, 1971). The basal cells lying beneath the centres of the cell columns are relatively inert. The reason for this arrangement is not known, although MacKenzie has speculated that a microgradient of chalone concentration may exist with its highest point beneath the centre of the overlying cell.

This arrangement has also led Potten (1974) to develop the hypothesis that each cell column, together with the underlying group of basal cells, forms a fairly independent 'proliferative unit'. He postulates that the basal cells beneath each column contribute post-mitotic cells mainly or only to that same column, although since most mitoses develop under the boundaries between the columns, and since the post-mitotic cells must move to whatever nearby column has an unfilled cavity at its base, this is most unlikely. There is in fact little or no evidence for this concept of epidermal structure, especially as so many epidermal types have a higher mitotic rate and so do not develop a columnar pattern.

4.3.2 *The roof-tile cell pattern*

When the number of new cells produced per day per cell column is about one, as it is in guinea-pig ear epidermis, the columnar pattern of cellular architecture is on the edge of disorganization. Only the slightest degree of hyperplasia, due perhaps to gentle ear scratching, is needed to change the situation.

The slightest degree of hyperplasia results in a small increase in the overall dimensions of the post-mitotic cells. Then, as the cells enter the base of the stratum compactum, their greater diameters make it impossible for them to fit into the columnar pattern. In theory these larger cells could form the same columnar pattern except that there would be fewer and wider columns per unit skin area. The alternative possibility, which in fact occurs, is that the larger cells will overlap each other like the tiles of a roof, the thicker central region of each flattened cell, where the nucleus still persists, being fitted against the thinner peripheral regions of the two or more cells above.

So far this overlapping pattern of cell stacking has only been described for guinea-pig ear epidermis (Bullough and Stolze, 1978), and again, once it has been established, it is evidently the configuration of the underside of the stratum compactum that acts as a template to guide into place the migrating cells of the stratum mobile.

4.3.3 *Cellular disorder*

With a further rise in the mitotic rate the roof-tile pattern is replaced by cellular disorder in the stratum compactum. This is evidently the normal condition of the epidermis of many mammals, including man; it is also seen in such natural forms of hyperplasia as mouse sole-of-foot. The larger numbers of new cells, passing more quickly across the wider stratum mobile, simply crowd haphazardly into the stratum compactum. There are too many of them arriving too quickly for the organization of any pattern to be possible.

This high rate of new cell production per unit skin area is caused in two distinct ways, which may operate separately or together. It may be the result of the increased mitotic rate alone, as already described, or it may be the result of a greater number of basal cells, which are consequently columnar, per unit area of baseline, a situation that is often combined with deeper baseline folding. Sometimes such basal cell crowding is alone responsible for the crowding of the cells traversing the stratum mobile.

Thus, for example, mouse sole-of-foot epidermis, which has a disordered pattern in the stratum compactum, has a mitotic rate not much greater than that of ear epidermis, which has a columnar pattern in the stratum compactum. However, as already noted, the number of basal cells per unit area of sole-of-foot epidermis is about five times that of ear epidermis owing to a combination of cell crowding and baseline folding.

4.3.4 *The stratum corneum*

The thickness of the dead stratum corneum parallels that of the living epidermis beneath. In normal epidermis it is thicker when the mitotic rate is higher, or when the basal cells are more crowded and more columnar in form, or when the baseline is more deeply folded, and these three factors are additive in their effects. Thus the stratum corneum of the mouse ear is thin, with about 8 stacked squames per column, while that of mouse sole-of-foot, with a slightly higher mitotic rate, greatly crowded columnar basal cells, and a folded baseline, is thick, with about 40 non-stacked squame layers.

In hypoplasia, as the epidermis shrinks into phase 1, not only does the living epidermis become no thinner as the mitotic rate falls, but so also does the stratum corneum become no thinner. Thus in pathological hypoplasia the squames must remain attached to each other for much longer than normal. The efficiency of the outer protective layer remains unimpaired.

4.4 Cellular re-organization after wounding

The changes that occur when unbroken epidermis becomes hyperplastic have been described above. However, the reconstruction of broken epidermis after wounding is a more complex process, which also provides further evidence of the capabilities of the cells in the three main epidermal layers.

When small areas of epidermis and superficial dermis have been removed, the first response is local inflammation and oedema, the dermis beneath the blood clot becoming swollen by capillary dilatation, collagen fibre swelling and the production of a watery exudate (Winter, 1972). Next the scab and the outer dermal region begin to dry, which causes the death by dehydration of the outermost cells of the decapitated hair follicles. These processes are complete after about 24 hours and the epidermis then begins to respond.

Wound closure is achieved primarily by epidermal cell migration. The great increase in the local mitotic rate merely provides the cells needed for the migration, as well as those subsequently needed to rebuild the epidermal thickness and to recreate the protective stratum corneum. The migrating cells are never also mitotic, and to avoid dehydration they travel through the moist connective tissue that lies beneath the dry outer layer. The new epidermal covering comes partly from cells that move inwards from the wound edges, but mainly (up to 90%) from cells that move outwards from the cut hair follicles. The speed of movement is about one cell diameter ($c. 7\mu m$) per hour, and in pig and man, together with the contribution from the cut hair follicles, an area of about $2.5\,cm^2$ is resurfaced in about 6 days.

Considering first the epidermis at the wound edges, the main question is the source of the migrating cells. Clearly they cannot come from the stratum compactum: not only are these cells firmly cemented together but they are also dying cells in the final phase of the ageing pathway, and so are unable to revert to mitosis. The migrating cells, which become mitotic when their movement ceases, must come from either the basal cell layer or from the lower part of the stratum mobile (that is before the cells have become expanded and plate-like) or from both.

The cells that migrate outwards from the broken hair follicles are said to come from the outer root sheath (Winter, 1972). This epithelium ensheaths the whole follicle but its cells above the level of the sebaceous gland duct have a markedly different appearance from those below that level. The upper cells look and behave like those of the surface epidermis with which they are continuous; the lower cells contain glycogen, are usually vacuolated, and only occasionally show any mitosis (Montagna and van Scott, 1958). It is not clear whether, according to the level of the cut, both these cell types can contribute to the new surface epidermis but certainly the upper cells, which are typical epidermal cells, can do so.

4.4.1 *The basal cell layer*

It has commonly been thought that wound closure is achieved by the migration of the adjacent undamaged basal cells, which simply slide sideways as a sheet across the dermal surface. In recent years this belief has had to be abandoned. In particular, Winter (1972) has shown that the epidermal cells do not move in this way, and indeed that the basal cells lateral to the wound do not move at all. These basal cells retain their normal characteristics: they remain firmly fixed to the dermis and they remain mitotic. Their only contribution to the closure of the wound is their exceptionally high rate of new cell production caused by the local fall in the G_1 chalone effectiveness.

4.4.2 *The supra-basal cell layers*

With the cells of both the stratum basale and the stratum compactum remaining immovable, the cells that migrate over the wound surface can only come from the stratum mobile. In describing their movement, Winter (1972) has emphasized that they do not crawl *en masse* as a cohesive sheet. Instead, the first supra-basal cell to move out drops straight down on to the baseline, where it rounds up to become the first of the basal cells of the new epidermis; the following cell then passes over it and drops down to become the second basal cell; and so the process continues by what can be called a 'cascade effect'. It seems that no single supra-basal cell moves more than two or three cell widths (i.e. $< 20\,\mu m$) from its original position. Thus the new basal layer is built up cell by cell, the time between the basal implantation of one cell and the next being about 1 hour. The movement ceases when the new basal cell layer coming from one direction meets that coming from another direction, that is when no hole remains for a supra-basal cell to drop into.

Each newly-formed basal cell then reverts to mitosis, so providing more new cells for movement laterally over the wound or vertically to form the new epidermis. Thus, according to Winter (1972), 'the slow advance of the epidermis over the wound is sustained by the production of cells in the new epidermis itself.' At any one moment there is an inner zone of moving cells, followed by a zone of high mitotic activity, which in turn is followed by a zone of declining mitotic activity as the new epidermis thickens and regains its normal chalone concentration.

It follows that, except in one particular, none of the epidermal cell types changes its nature. Wound healing is achieved merely by an accentuation of the processes that occur in normal epidermis, with higher mitotic activity basally, more active cell movement supra-basally, and faster keratinization distally. The one change is that the supra-basal cells revert to the mitotic cycle when they come into dermal contact and the action of the mesenchymal factor again predominates. Until this happens the moving cells never divide, partly because they are post-mitotic and it takes some time to reactivate the mitosis operon, and partly because cell movement makes such heavy energy demands that not enough remains to support mitosis.

4.5 Summary and conclusions

From a dynamic point of view the epidermis consists of a stratum basale, the mitotic-cycle cells of which are gripped by the dermal surface, a stratum mobile, the post-mitotic cells of which move individually along specific migration paths, and a stratum compactum, the dying cells of which are firmly cemented together in one of several possible patterns.

4.5.1 Local epidermal thickness

While the general thickness of the epidermis is determined by the value of the R ratio, which is evidently species-specific, and which may be a function of the value of the $G_1 : G_2$ chalone concentration ratio, the local differences in epidermal thickness that are characteristic of particular body regions are determined by local differences in the mitotic rate, in the strength of the dermal grip on the intermitotic G_{1b} cells, and in the degree of folding of the dermo-epidermal junction. These three factors, individually or in combination, determine the number of basal cells per unit skin area. Then, according to the N ratio, the greater the number of basal cells, the greater the number of distal cells, and the thicker the epidermis.

Furthermore, it is probable that all these three factors are dermis-dependent. The local epidermal mitotic rate is proportional to the local concentration of the mesenchymal factor, the degree of basal cell crowding is proportional to the strength of grip exerted by the dermal surface, and the degree of the baseline folding, though its cause is more obscure, seems also to be determined by dermal influences.

The general conclusion is that the epidermis is dependent, first, for its origin on dermal induction (5.1), second, for its continuing survival on the mitotic stimulus of the mesenchymal factor, and third, for its regional peculiarities on other localized dermal influences.

4.5.2 Intra-epidermal tension

The tension that exists within the basal cell layer is obvious from the crowded columnar structure seen in many types of epidermis, and from the way in which the intermitotic cells are forced out by the pressure exerted by an adjacent mitosis. The basic cause is the grip of the dermis on the basal cells, together with the attachment of these cells to each other, acting against the lateral pressure built up by mitotic activity. The stronger the basal grip, the greater the lateral tension on the basal cells, and the more compressed and columnar they become. Then, the more columnar they are, the smaller the area of contact between each cell base and the dermal surface, and the weaker the dermal grip. The general lateral tension within the basal cell layer is a function of these various opposing factors.

This general tension is then self-maintaining at a level just below that needed to cause the extrusion of the intermitotic cells, which have the weakest basal attachment. Any extra local tension arising from the development of a mitosis is immediately counteracted by the extrusion of a local intermitotic cell.

Of considerable practical importance is the fact that, when the epidermal mitotic rate rises rapidly to a high level after tissue damage, at least three mechanisms come into operation to prevent any increase in tension that might otherwise disrupt the basal cell layer. Epidermal structure thus remains undistorted, in contrast to what happens for instance in epidermal carcinomata. These mechanisms are: the extrusion from the basal layer of intermitotic G_{1b} cells for as long as these cells remain available; the increasing incidence of vertically angled mitoses as the G_{1b} cells become fewer in number; and, in many types of epidermis, the deeper folding of the dermo-epidermal junction to accommodate the extra mitotic-cycle cells. It is important to emphasize that these mechanisms operate in natural hyperplasia, such as mouse sole-of-foot epidermis, as well as in pathological hyperplasia.

It may be added that some speculation exists that basal cell extrusion may be caused not by cell population pressure, as described above, but either by the pulling out of the basal cells by the supra-basal cells or by the unpressurized distal migration of the older basal cells (Etoh *et al*, 1975; Skerrow, 1978). There is no good evidence for these propositions (Mitrani, 1978).

4.5.3 *The fate of the basal cells*

It is an old and common misconception that all basal mitoses occur in a vertical direction, so that one daughter cell remains on the baseline to divide again while the other is supra-basal and post-mitotic. From this arises the further misconception that the two daughter cells are inherently different *ab initio*, the one being pre-programmed for the next mitosis and the other for tissue function, which in epidermis is keratinization. In consequence, this type of mitosis has been termed 'unequal' or 'asymmetric', and it has been postulated that 'the asymmetric distribution of cytoplasmic determinants at cell division' accounts for the different fates of the two daughter cells (Gurdon, 1977).

It cannot be too strongly emphasized that in the adult epidermis, and probably in all the other epithelia as well, the two daughter cells are identical in their potentialities, that they both tend to remain basal, and that the fate of any epidermal cell depends on the environment in which it finds itself.

This conclusion is also supported by the probability theory of mitosis (2.5.3). Just as the entry of a basal cell into mitosis is a random event, so in consequence must the extrusion of a closely adjacent intermitotic G_{1b} cell be a random event. The extruded cell would have remained a mitotic-cycle cell had it remained basal; being extruded its post-mitotic synthesis of keratin is dictated by the chemical messengers in its supra-basal environment.

4.5.4 *Supra-basal cell movement*

The factors that direct the movements of the cells of the stratum mobile are unknown. All that is known is that if these movements take place slowly the result is the formation of a columnar cell pattern; if they proceed more quickly the result is a roof-tile pattern; and if they are fast no pattern is formed. Evidently, if a pattern is to be built up, there must be adequate time for the cells to respond to the directing factors.

These factors may be physical, as when the undersurface of the stratum compactum may act as a template, but they may also be humoral. In a review of cell movement, Abercrombie (1980) has concluded that 'cells are extremely sensitive in their locomotion to their surroundings, to other cells with which they come in contact, to the substratum on which they crawl, and to the liquid medium that bathes them'. He lists three main influences that may direct the movements of migrating cells. These are, first, that cells tend to move from a region of high cell density to one of low cell density; second, that they are guided on their way by the nature, and especially the shape of the substratum over which they move; and third, that they are influenced by chemotaxis in such a way that they tend to move along the line of maximum steepness of a chemical concentration gradient. All these factors could play some part in the movements of the stratum mobile cells.

One practical problem arises from the fact that the way the cells move to create one squame pattern or another seems to have little or no effect on the functional ability of the protective stratum corneum. The possibility has been suggested that an organized columnar structure may offer more protection when the stratum corneum is at its thinnest in phase 1 type epidermis, while in thicker phase 2 type epidermis the mere bulk of the stratum corneum may in itself afford an adequate protection.

Thus the significance of these various forms of epidermal cellular architecture remains a matter for speculation. However, in the formation of such complex epidermal derivatives as claws, scales, and hairs, the factors that determine the specific migration paths of the post-mitotic cells are of obvious importance. Billingham and Silvers (1968) have stressed the complexity of the patterns of cell movements that must exist in such cases, and have concluded that the morphogenic stimuli must come mainly from the underlying mesenchyme.

Chapter 5

The dominance of the dermis

5.1	*Dermis-induced differentiation*		75
	5.1.1	*In the embryo*	75
	5.1.2	*In the fetus and neonate*	76
	5.1.3	*In the adult*	76
5.2	*Dermis-induced modulation*		78
	5.2.1	*In adult epidermis*	78
5.3	*Summary and conclusions*		79

Although to a great extent the epidermal cells control their own fate through their chalone control mechanism, it is clear that they are also dependent in many ways on influences emanating from the dermis. In adult epidermis the chalone mechanism cannot function without the pro-mitotic action of the mesenchymal factor, and the regional differences in epidermal thickness and structure are dermis-dependent. However, the influence of the dermis is more wide-ranging than this. The epidermis owes to the embryonic and fetal dermis its differentiation as a keratin-synthesizing epithelium, as well as its ability to give rise to a range of derivative tissues such as mammary glands, sweat glands, and hair follicles.

These are typical induction processes, which means that they depend on the activation of certain gene operons and the inactivation of others. This type of dermo-epidermal interaction may be considered first, leaving the other type of interaction that does not involve changes in gene responses, and that has been called modulation, to be considered later (5.2).

5.1 Dermis-induced differentiation

In an embryo it is a general principle that the differentiation and morpho-genesis of any epithelial tissue is dependent on the influence of the adjacent mesenchyme, which in the case of the epidermis is the dermal mesenchyme. The evidence for this statement comes from studies of mesenchymal-epithelial relations in a wide range of different epithelial tissues, mainly from the embryos of birds and mammals (see Wessells, 1977).

5.1.1. *In the embryo*

The available evidence is so extensive that only the briefest summary can be given. From experiments on a range of different epithelial and mesen-chymal tissues it has been found: that an isolated epithelium will never differentiate; that when it is recombined with its usual mesenchyme it will differentiate normally, even if this mesenchyme comes from a different species of mammal; and that, at least at an early enough stage in the differentiation process, when an epithelium is combined with an alien mesenchyme, this mesenchyme is dominant so that the epithelium differ-entiates as would the epithelium that is normally associated with that mesenchyme.

Thus, for example, when the undifferentiated epidermis of a bird embryo is combined with mammary gland mesenchyme, it forms a typical branching tubular mammary epithelium, and similarly when un-differentiated epidermis from embryonic sole-of-foot is combined with tooth organ mesenchyme it differentiates to form a tooth.

From these and other similar examples, Slavkin (1980) has concluded 'that mesenchyme produces a specific set of molecular instructions which determine the cytodifferentiation of the associated epithelium'.

However, the range of syntheses that can be initiated in any epithelial cell, including epidermis, becomes progressively more restricted during embryogenesis, and such remarkable inductions as those described above become less and less possible. Each epithelial cell type assumes organiza-tional capacities of its own so that, as one example, when pancreatic epithelium has been induced by its mesenchyme but is not yet histolog-ically differentiated, such differentiation will occur normally when the epithelium is combined with an alien mesenchyme. At this later stage the mesenchyme is merely needed to provide the usual mitotic stimulus, which is clearly of a different nature from that of the original induction.

5.1.2 *In the fetus and neonate*

After epidermis has become semi-differentiated, as in the human fetus or the fetal and neonatal mouse, it still retains the capacity to give rise to a range of other tissues including especially hair follicles, sebaceous glands and mammary glands. As usual, the inducing signals come from the underlying dermis but the responses of the epidermal cells are strictly limited to what they remain capable of at that late stage in development. As one extreme example, when bird epidermis is combined with mammalian dermis it is induced to produce feathers and not hair, and conversely when mammalian epidermis is combined with avian dermis it is induced to produce hair and not feathers (Slavkin, 1980).

In both birds and mammals the regional differences in feather and hair types are also dermis-dependent. In birds, epidermis transplantation experiments have shown that the small body feathers and the large wing feathers are induced by the body dermis and the wing dermis respectively, while the dermis of the leg induces the production of scales. Similar experiments with mammals have shown that the regional hair types are also induced by the regional dermis types.

It is also significant that such epidermis-derived tissues as the hair follicles develop according to a pattern, whereby they are roughly evenly spaced across the surface of the skin. As the skin surface rapidly expands, secondary and then tertiary hair follicles are formed in the widening spaces that develop between the primary follicles. The impression is given that each follicle creates around itself a field of influence within which the development of any other follicle is inhibited. An inhibiting influence of this type could also be based within the dermis through the production of inhibitory messenger molecules in each dermal papilla.

5.1.3 *In the adult*

After the fetal and neonatal periods, when all the hair follicles and other epidermis-derived tissues have been formed, the epidermis itself can be regarded as fully differentiated. It then begins to respond to the epidermal G_1 chalone (2.3.4). However, the evidence shows that, in greater or lesser degree, some of the inducing influences of the various dermal regions are still operating, most notably that which activates the epidermal mitosis operon.

Furthermore, some specialized derivative epithelia, such as those of the mouth and oesophagus, can sometimes be induced to respond to the signals produced by an alien type of dermis; indeed the evidence is that any failure to respond may lie not so much in the epithelial cells as in the weakened inductive powers of the dermis. Thus in the hamster, the lingual, oesophageal, and cheek pouch epithelia all retain their original characteristics when transplanted on to ear dermis, but all transform into typical epidermis when transplanted on to trunk dermis (Billingham and Silvers, 1968). The conclusion is that the epithelia are capable of responding but that only in the trunk region is the dermal influence strong enough to induce them to do so.

A similar but even more dramatic example of an adult epidermis-derived tissue responding to dermal induction is seen when, in the rabbit, the mammary gland epithelium is transplanted on to the dermal bed of an open skin wound. Billingham and Silvers (1968) have described how this epithelium established and transformed itself until 'the entire wound was completely resurfaced by a heavily keratinized hyperplastic epidermis of mammary gland origin'.

Further information comes from studies of hair follicle regeneration. Adult mammalian epidermis is usually unable to regenerate any hair follicles that are lost, for instance by skin damage, and again the evidence is that this is basically due to the weakened inductive power of the dermis. Studies in deer have shown that every spring each regenerating antler is covered by a rapidly expanding epidermis from which innumerable new hair follicles are formed (Goss, 1972). It has been found that this new skin originates from the region immediately around the antler base, and that the rest of the body skin is, as usual, incapable of new hair follicle production. However, when the epidermis of the antler base is removed and replaced by body epidermis, then during antler growth this transplanted epidermis expands and produces new hair follicles on the usual grand scale. The general body epidermis has not lost its competence; the general body dermis has lost its inductive power. A fetal-type dermis has been retained at the antler base to serve an adult need.

Experiments of these kinds have their obvious limits. It may be possible to induce epidermis to form a variety of epidermis-derived tissues, or to induce such tissues to form epidermis, but the cells of the genus epidermis are limited to epidermis-type syntheses. Consequently, when adult epidermis is transplanted on to the connective tissue of the kidney surface it can only continue to form the usual epidermal type of epithelium (Billingham and Silvers, 1968).

5.2 Dermis-induced modulation

It is clear that the normal steps in the differentiation of the epidermis and its derived tissues, described above, are dependent on induced changes in the pattern of gene activity, and it is probable that the various experimental modifications of epidermal structure that are also described are due to the same cause. They are true examples of differentiation.

However, the other controlling actions exerted by the dermis on the epidermis may not involve the activation and inactivation of genes, and may therefore be better called modulations. There is certainly a significant difference between, for instance, the differentiation of epidermal cells into hair follicle cells and the modulation of normal epidermis into a thicker regional type of epidermis. In this latter case the only change in gene activity may be an increase in the rate at which RNA is synthesized. However, it must be emphasized that the boundary between differentiation and modulation is a vague one, which is mainly due to our lack of knowledge of the precise nature of the various epidermal responses.

5.2.1 *In adult epidermis*

All basal epidermal cells are equipotential, in that all are competent to modulate their behaviour in response to whatever dermal stimuli they receive, whether naturally or after epidermal transplantation. Thin ear epidermis transplanted on to sole-of-foot dermis produces a thick compact epithelium that is indistinguishable from normal sole epidermis, while conversely, thick sole epidermis implanted on to ear dermis is transformed into thin ear-type epidermis.

From this and similar experiments, Billingham and Silvers (1968) concluded that the maintenance of epidermal structure 'turns upon the persistent influence of the underlying dermis', and that each regional epidermal type 'is dependent upon the continuous action of specific, regionally distinctive, inductive stimuli derived from the local populations of dermal fibroblasts'. The nature of some of these stimuli is now becoming apparent. In particular, dermal influences determine epidermal thickness and structure through their stimulus to epidermal mitosis, through the strength of the dermal grip on the basal epidermal cells, and through the degree of folding of the dermal surface.

These influences must be constantly maintained or the epidermis may revert to the simple structure typical of the body surface. For instance, when corneal, conjunctival and various other forms of epidermal epithelia are maintained in the same *in vitro* conditions they all give rise to the same type of epithelium, so that Sun and Green (1977) have concluded that 'much of the distinctive phenotype of these epithelia *in vivo* must be due to external modulation, and relatively little to permanent intrinsic divergence during development'.

It is, therefore, the dermal cells and not the epidermal cells that are regionally distinct, and these distinctions remain stable throughout life. Each epidermal region, and therefore each dermal region, always remains sharply demarcated. Thus, in the mouse, the sharp boundary between the thick sole-of-foot epidermis and the thin side-of-foot epidermis does not become blurred with time, that is with the continuous replacement of the cells on both sides of the boundary. Either there is never any sideways wandering of these cells, even after wounding, or more probably, when such wandering across the boundary does occur, the invading cells are immediately caused to adopt the characteristics proper to their new position, as happens after epidermal transplantation.

After wounding or other type of skin damage both epidermis and dermis become hyperplastic. If the damage is not too severe the dermis continues to issue its normal instructions to the newly forming epidermis, so that it develops its appropriate regional structure. Only if the dermis has been badly damaged, as by a deep burn, may the instructions to the epidermis be so distorted or inadequate that the new epidermal structure is abnormal.

5.3   Summary and conclusions

The influence of the dermis on the epidermis is obviously all-pervading. The active presence of the dermis is essential both for the origin of the epidermis and for its continuing existence throughout life; it is similarly essential for the origin and continuing existence of all the epidermis-derived tissues.

Its actions can be summarized as follows: in the embryo the dermal mesenchyme induces the differentiation of a proto-epidermis, that is of an epithelium of multipotential stem cells that are capable of a limited range of syntheses, especially of keratin; in the fetus the dermis induces the differentiation, from the proto-epidermis, of a range of localized and specialized glandular tissues, which secrete, for instance, milk (following appropriate hormonal stimulation), sweat, sebum or from the hair follicles long strands of tough keratin; also in the fetus, by a pattern of inhibitory fields, the dermal papillae may ensure the proper spacing of the hair follicles over the body surface; at the same time, the dermis induces the terminal differentiation of the epidermis itself; the dermal cells synthesize the mesenchymal factor which activates the mitosis operon in the adjacent epidermal cells; and finally, through the power of the mesenchymal factor stimulus combined with the strength of the dermal grip on the basal epidermal cells and the degree of folding of the dermal surface, the dermis modulates the thickness and the cellular architecture of the overlying epidermis.

Each of the many steps in cellular differentiation that give rise to the adult epidermis and to the various epidermis-derived tissues also results in the synthesis of tissue-specific G_1 and G_2 chalones and of their matching chalone receptor sites. In all the chalone mechanisms so formed the mesenchymal factor plays an essential role in the maintenance of cellular homeostasis. Without its stimulus to mitotic activity in the basal cell layer there would be no cells for the G_1 and G_2 chalones to control.

This raises a problem. Since cellular homeostasis in an epithelial tissue such as epidermis cannot be maintained without the mitotic stimulus to its basal cells coming from the adjacent dermis, it would seem that cellular homeostasis in a connective tissue such as the dermis, in which no similar basal cell layer exists, could not be achieved in the same way. Some different mechanism for cellular homeostasis must be required; this problem is considered in Chapter 8.

Homeostasis in Other Tissues

The cells combine to construct the tissues, their metamorphoses being mysteriously governed by their plastic nature and by a mode of force operating unconsciously upon the matter, but according to a law of order and harmony.

Richard Owen, 1804–1892

Chapter 6

The epithelial tissues

6.1	The messenger molecules	84
	6.1.1 The mesenchymal factor	84
	6.1.2 The chalone systems	86
	6.1.3 The chemical nature of the chalones	87
6.2	Cellular homeostasis	88
	6.2.1 The ageing pathway	89
	6.2.2 Wound healing	90
	6.2.3 Compensatory hypertrophy	90
	6.2.4 Periodic tissue hypertrophy	92
6.3	Tissue organization	95
	6.3.1 Epidermis-derived tissues	95
	6.3.2 Other epithelial tissues	96
6.4	Summary and conclusions	97
	6.4.1 Cell gain and cell loss	97
	6.4.2 Tissue structure	98
	6.4.3 Tissue mass	98
	6.4.4 The structure of a chalone mechanism	99

The way in which epidermal homeostasis is continually maintained by cellular responses to gradients of messenger molecules has been outlined, although the details of what is evidently a complex mechanism are not well enough known for final conclusions to be drawn. The question that now arises is whether this same type of chalone mechanism also exists in the other body tissues. In support of this thesis it is well known that in all mitotic epithelial tissues there is a perfect balance between the production of new cells, whether slow or fast, and the loss of old mature cells; that after wounding the healing response involves the same two processes of cell migration and cell division as in epidermis; and that these processes occur only in the tissue that has been damaged, and not in the adjacent tissues, so that the control mechanism must be tissue specific.

Certainly it would be surprising if the basic homeostatic control mechanisms of the various epithelial tissues proved to be markedly different from each other, except in the one important detail that each must depend on a tissue-specific element, or chalone, that is the key to the operation of the mechanism.

6.1 The messenger molecules

One thing is certain, namely that the processes underlying the mitotic cycle and the post-mitotic ageing pathway must be the same in all tissues that are capable of cell replacement. The only questions are whether the three types of messenger molecule, the non-tissue-specific mesenchymal factor and the tissue-specific G_1 and G_2 chalones are present and active in all epithelial tissues, and whether, as in the epidermis, the two stress hormones are involved.

6.1.1 *The mesenchymal factor*

This factor and its pro-mitotic action have mostly been studied in relation to the newly-formed tissues and organs in the embryos of birds and mammals, this being part of the extensive research that has been carried out into the complex mesenchymal–epithelial induction phenomena (see Wessells, 1977). For present purposes the general conclusion is that in all embryonic epithelial tissues so far studied mitotic activity is confined to those cells that are adjacent or close to the mesenchyme, whether it is the normal mesenchyme or that taken from some other organ. This relationship has been demonstrated in the epithelia of the gizzard, salivary gland, mammary gland, and pancreas, while in the lung epithelium the mitotic rate is directly proportional to the amount of mesenchyme present. The experimental evidence indicates that the embryonic mesenchyme produces a messenger molecule that stimulates mitotic activity in all types of epithelia, which are themselves unable to produce it. To this can be added that to promote mitotic activity in embryonic epithelia *in vitro* it is necessary to add fresh 'embryo extract', which evidently contains the stimulating factor.

The evidence obtained from the fewer adult epithelial tissues that have also been studied shows that the mesenchymal factor is present and active throughout the whole of life. Besides adult epidermis, this evidence comes particularly from studies of human prostate epithelium, which *in vitro* without its connective tissue fails to grow and loses its testosterone-responsiveness (Franks, 1963; Franks *et al*, 1970), and from bladder epithelium, which without its connective tissue degenerates and dies, whereas with this tissue it can be maintained *in vitro* (Hodges, 1967). Again it is relevant to note that epithelia in general may be stimulated to mitotic activity *in vitro* by the addition of fresh serum as a substitute for the connective tissue. As already shown, such serum contains a mitosis-promoting glycoprotein of some 60 000 daltons, which could perhaps be the mesenchymal factor emanating from the various connective tissues of the body (Levine *et al*, 1973; Houck *et al*, 1973).

Besides these experimental results there is much descriptive evidence to indicate that epithelial mitotic activity is connective-tissue-adjacent. This is particularly obvious in the skin. Just as the mitotic activity of the epidermis is strictly dermis-adjacent, so also is that of the sebaceous glands and of the eccrine sweat gland ducts. It is even true of the intra-epidermal melanocyte cell population and probably also of the Langerhans cell population.

The active hair root is particularly interesting in that its mitotic activity is confined to those cells that lie adjacent to a specialized region of the dermis called the dermal papilla. This situation can be compared with that in the intestinal crypt epithelium, where mitosis is confined to those cells that lie within the deeper connective tissue layers. One suggestion has been that such special connective tissue regions produce qualitatively different types of mesenchymal factor, but it is more probable that the differences are quantitative (Bullough, 1975). In support of this suggestion it can be seen that the cell concentration within the dermal papilla is particularly high, and it is known that when the dermal papilla is transplanted to a position close beneath the epidermis the adjacent epidermal cells respond with an abnormally high rate of mitosis.

The general conclusion is that the mesenchymal factor, produced throughout life by the various kinds of connective tissue, is non-tissue-specific in its mitosis-promoting action on the various epithelia, and that within each epithelium it is present in effective concentration (relative to the mitosis-inhibiting chalone) only around those cells that are in connective tissue contact. The evidence is that it is present in the intercellular spaces and 'that it functions at the cell surface in promoting mitosis'

(Wessells, 1977). Furthermore, with so much connective tissue present in the body, enough mesenchymal factor may escape into the blood not only to explain the serum action on mitotic activity *in vitro*, but possibly also to promote capillary-adjacent mitotic activity in some epithelial tissues *in vivo*. There is also some evidence suggesting that the mesenchymal factor may be produced by the blood cells themselves (Ristow, 1982).

6.1.2 *The chalone systems*

Information on the chalone systems of the various body tissues is accumulating rapidly (see Bullough, 1975; Iversen, 1981), and the main conclusion emerging is that each tissue responds only to its own specific chalone. Studies of the tissues of the skin and of the blood (8.2) have been especially revealing. Thus in the skin, in addition to the epidermal chalone system, it has been found that the various epidermal derivatives, the sebaceous glands, the eccrine gland ducts, the mammary gland ducts, and also evidently the hair follicles, all possess their own specific chalone systems. Within the epidermis the melanocyte cells are also separately controlled, as are the cells of the underlying dermis (8.1). Thus within one organ there exist many separate chalone systems which allow each cell type to respond independently to the demands made on it.

From this the impression is gained that the body must contain a multitude of separate chalone control mechanisms. This may well prove to be true but there is also some evidence that different tissues may share the same chalone mechanism. One example, already noted, concerns the epidermal chalone system which is present and active against mitosis in the lens epithelium and the oral and oesophageal epithelia, and more remarkably which is also present in the non-keratinizing epithelia of the trachea and the bronchioles. The epidermal chalone also strongly inhibits mitosis in a rapidly growing bronchiolar carcinoma (Iversen, 1981).

Thus the evidence is that, while some tissues which are closely related embryologically possess their own specific chalone control mechanisms, other tissues which are not so closely related may share the same chalone mechanism (10.2.1).

Besides the chalone systems of the skin there is much evidence, of greater or lesser reliability, for the existence of a number of distinct chalone systems in the various epithelia of the alimentary canal and in their derivative tissues (Iversen, 1981). In addition to the oesophageal epithelium, controlled by the epidermal chalone, there is some evidence for a specific chalone system in the gastric epithelium and strong evidence for specific chalone systems in the crypts of the small and large intestines. The small intestine contains both G_1 and G_2 chalones which are active against both embryonic and adult epithelia (Brugal, 1976; Sassier and Bergeron, 1978). The colon contains a G_1 chalone which has been partly purified and which has an apparent molecular weight of $10\,000$–$50\,000$ daltons (but see 6.1.3). This chalone is also active against colon carcinoma cells (Kanagalingam and Houck, 1976).

The liver presents a particularly interesting and complex picture, but unfortunately much of the voluminous literature is of doubtful significance (Bullough, 1965). The first clear evidence for the existence of a liver chalone was that of Saetren (1956) and since then much information has accumulated indicating the existence of both G_1 and G_2 chalones. All this work has related specifically to the hepatocytes; the other liver tissues presumably possess their own chalone control mechanisms but much more remains to be done before a clear picture can be obtained.

In addition to all these chalone systems, others have been described in a number of heterogeneous tissues: the kidney tubules contain both G_1 ($MW = c.$ 5000 daltons) and G_2 chalones; the lung alveolar cells are evidently controlled by a chalone mechanism that is distinct from the bronchiolar (epidermal) mechanism; the testis, or more specifically the spermatogonia, are believed to produce their own G_1 and G_2 chalones, as also do the seminal vesicle and the prostate; and there is strong evidence for the existence of an endothelial cell chalone (Iversen, 1981).

Finally, it can also be mentioned that, in the non-epithelial tissues, there is evidence for a smooth muscle chalone (7.1.1) and overwhelming evidence for a variety of chalone systems associated with the structural connective tissues (8.1) and with the various blood cell tissues (8.2).

The conclusion must be that tissue-specific chalones are universally present, at least in those tissues that retain a mitotic potential.

6.1.3   *The chemical nature of the chalones*

The multiplicity of tissue-specific chalone mechanisms within the body, all controlling the non-tissue-specific processes of mitosis and post-mitotic cell ageing, suggests that the chalones themselves will prove to be a family of closely related molecules. The earlier attempts to understand the chemical nature of the G_1 and G_2 chalones led to a belief that they may be glycoproteins of between some $50\,000$ to $> 100\,000$ daltons. On theoretical grounds it was also thought that these molecules would prove to be unstable and short-lived since only in these circumstances would the cells be able to respond rapidly to changing conditions. Both these propositions now seem to be mistaken.

Regarding the G2 chalones there is still little information, and what there is relates only to the epidermal chalone (see 2.4.4; and Isaksson-Forsen *et al*, 1977; Okulov *et al*, 1978). Much more information exists regarding the G1 chalones, especially those of the epidermis, lymphocytes, granulocytes, and fibroblasts (Iversen, 1981). The original estimates of molecular weight were mostly high, up to 40 000 daltons in the case of the epidermis, but the latest information, reviewed by Patt and Houck (1980), has shown that the molecular weights are much smaller than had been believed. The lymphocyte and granulocyte chalones are now thought to be stable peptides of some 600–700 daltons or less, while the epidermal and fibrocyte chalones are estimated at < 10 000 daltons. Except for the granulocyte and erythrocyte chalones they bind avidly with a range of larger molecules, and consequently it has been suggested that *in vivo* they may normally be coupled to carrier molecules, which may be glycoproteins. The latest information on the granulocyte chalone indicates that it is a pentapeptide (Paukovits and Laerum, 1982; see 8.2.3).

In all the research on chalone purification, chalone structure and chalone action little attention has been given to the question of possible stress hormone involvement. However, Bullough (1965) has reviewed the extensive evidence from a range of tissues for the common existence of diurnal mitotic cycles, for mitotic inhibition in stressful situations, as well as for the anti-mitotic actions of adrenalin and of the glucocorticoid hormone. All this evidence, although indirect, suggests that, as in the epidermis, the actions of adrenalin and the glucocorticoid hormone impinge on most of the various chalone mechanisms to strengthen their anti-mitotic action.

6.2 Cellular homeostasis

If G1 and G2 chalones, acting as they do in epidermis, are to be found in most, if not all, tissues of the body, then the mechanism of cellular homeostasis must also be universally similar. In particular there must always be a close connection between the rate of cell gain by mitosis (limited by the G1 chalone) and the rate of post-mitotic cell ageing (possibly limited by the G2 chalone), since it is fundamental to cellular homeostasis that any change in the mitotic rate is immediately matched by an equivalent change in the rate of cell ageing and death.

6.2.1 *The ageing pathway*

Furthermore, in all tissues the phases of the ageing pathway must be the same: first, there is the A1 phase when tissue function is beginning, or has begun, but when in the crisis of tissue damage the mitotic genes are capable of reactivation; second, there is the A2 phase when the cells are mature and functional and the mitosis genes are finally silenced; and last, there is the D phase when all the genes are silenced, the nucleus is dead or absent, and limited cellular survival depends on preformed mRNA and enzymes.

The evidence suggests that this is indeed the common pattern and that the epithelial tissues differ from each other only in the speed at which the post-mitotic cells traverse the ageing pathway. When the mitotic rate is high, as in the duodenal mucosa, a large proportion of the total cell population is in the mitotic cycle, which means that the post-mitotic cells must age and die quickly (in *c.* 24 hours); when the mitotic rate is moderate, as in the sebaceous glands, a larger proportion of the cell population is in the ageing pathway in which they remain longer (*c.* 21 days); and when the mitotic rate is negligible, as in the hepatocytes, most, if not all, the cell population remains almost static in the A1 and possibly also A2 phases (but not in the D phase). In tissues in which mitoses are rare or absent the post-mitotic functional cells may have a life expectancy that hypothetically is longer than that of the animal itself (Chapter 7).

Such evidence shows that the ratio, rate of cell gain : rate of cell loss, is constant in all tissues. Other evidence confirms that when in any tissue the mitotic rate changes, so in equal degree does the rate of cell ageing, and, therefore, of cell loss. In the sebaceous glands a four-fold reduction in the mitotic rate is accompanied by a four-fold increase in the duration of the ageing pathway (Bullough and Ebling, 1952); in the vagina and uterus, during an oestrogen-induced hyperplasia, the raised epithelial mitotic rate is matched by the earlier death of the post-mitotic cells (the post-mitotic life span is reduced from several weeks to only 1–2 days); in the bladder epithelium a stimulus to the normally low mitotic rate is matched by a shortening of the ageing pathway from about 1 year to about 1 week (Stewart *et al*, 1980); and in rat liver hepatocytes the stimulus to mitosis in cirrhosis is accompanied by a reduction in the ageing pathway from an estimated 450 days to about 26 days (MacDonald, 1961).

Furthermore, as in epidermis, no tissue can disappear because of a reduced mitotic rate, as during chronic stress, and no tissue can explode into uncontrolled growth during hyperplasia caused by chronic damage or, in hormone-dependent tissues, by chronic hormone excess. Thus the rules governing phase 1 type hypoplasia (3.1.2) and phase 2 type hyperplasia (3.1.3), including wound healing, must also be universal.

6.2.2 *Wound healing*

The repair of damage to epithelial tissues follows the same sequence as that seen in epidermis: first, there is the migration into the wound cavity of closely adjacent undamaged cells, which may be in the post-mitotic A_1 phase; and second, there is the sharp rise in the mitotic rate in those cells that are not involved in migration and that were previously either in the mitotic cycle or in the A_1 phase of the ageing pathway. The control of cell migration is not yet understood; the raised mitotic rate may be due both to the reduced sensitivity of the cells to the G_1 chalone and to the locally reduced concentration of this chalone.

In all wounds that heal normally the extra numbers of new cells formed are retained because the healing of each region of the wound is completed well within the time limit set by the reduced life span of these cells.

6.2.3 *Compensatory hypertrophy*

The repair of a wound involves only a local response, but in certain massive tissues and organs the loss of cells, if it is great enough, can lead to a general mitotic response throughout the whole remaining cell mass, which thereby recovers its normal bulk. The best known examples of such compensatory hypertrophy include: the growth of the remaining kidney after unilateral nephrectomy; the mitotic response that develops not only in the remaining liver fragment after partial hepatectomy but also in a distant liver tissue implant or in the unaffected liver of a parabiotic twin; and the increased rate of blood cell production in the bone marrow after severe haemorrhage (Bullough, 1965, 1975). Such examples indicate that, at least in these tissues and organs, the mitosis-controlling chalones not only act locally but can pass throughout the whole body, evidently in the blood and lymph. The epidermal G_2 chalone has been found in urine (Bullough and Laurence, 1971, unpublished data) and this may be the ultimate fate of many chalone molecules.

The response of the hepatocytes to liver damage can be summarized as follows: after the loss of up to about 10% of the total liver mass the mitotic response remains local as in any typical wound; after the loss of more than 10% of the liver mass an increased mitotic rate is seen throughout the whole of the remaining liver tissue, the degree of increase in the mitotic rate being in direct proportion to the percentage of liver tissue lost (Bucher, 1963; Bucher and Swaffield, 1964). The responding hepatocytes must have been in the post-mitotic A_1 phase in which, although fully functional, they can revert to mitosis whenever the chalone concentration falls below a certain critical level. Experiments have shown that this mitotic response can be prevented by injections of liver extract equivalent to the amount of liver tissue that was removed (Saetren, 1956). Similarly, the mitotic response in a single remaining kidney can be inhibited by injections of the extract of a whole kidney, and the mitotic response of each type of blood cell can be tissue-specifically inhibited by semi-purified extracts of the relevant G_1 chalone (Rytömaa, 1976a; Iversen, 1981).

Attempts have been made to explain compensatory hypertrophy in terms of the heavier metabolic demand that is made on the remaining undamaged tissue (see Bullough, 1967), a hypothesis that evidently derives from the well-known increase in mass of the skeletal muscles after extra exercise. However, the mitotic response of a mitotic tissue cannot be compared with the response of a tissue that is incapable of mitosis. A non-mitotic skeletal muscle grows by the increase in mass of each of the exercised muscle cells (7.1.2).

Most of the experiments that have been designed to show that extra metabolic work results in increased mitotic activity have involved the liver, which has been challenged to detoxicate excessive amounts of poisonous substances. Where such experiments have apparently succeeded the results may be ascribed to cell damage, such as occurs in human liver cirrhosis, which has the same effect as does partial hepatectomy.

Although other factors may sometimes be involved, compensatory hypertrophy is basically achieved through a simple negative feedback mechanism. In all tissues it seems that a higher intracellular chalone concentration is in balance with a lower intercellular chalone concentration. In the case of massive tissues, such as the liver hepatocytes which produce relatively large amounts of chalone, the intercellular chalone concentration is also in balance with a still lower, but nevertheless significant, systemic chalone concentration. When, after large-scale loss, there is a reduction in the amount of chalone synthesized, there must also be an appreciable fall in the blood chalone concentration. This in turn, by increasing the concentration gradient between the tissue and the blood, must lead to a balancing fall in both the inter- and intracellular chalone concentrations in the remaining undamaged tissue, including any similar tissue at a distance in the body or in a parabiotic twin. The consequent degree of increase in the mitotic rate will then, of course, depend on the proportion of the original tissue mass that has been lost, as is observed.

When the tissue has grown back to its normal mass relative to the total body mass, both the systemic chalone concentration and the tissue chalone concentration will rise to their proper balanced levels, and mitotic activity will return to normal. Compensatory hypertrophy will be complete.

The reason why in these circumstances a mass of new cells is gained without any equivalent loss of old cells has already been explained (3.1.3). During compensatory hypertrophy in the liver of a rat the life span of the newly-formed post-mitotic cells is only about 26 days, which compares with the normal figure of about 450 days, but this shorter period is more than adequate for the complete regeneration of even a heavily damaged liver (MacDonald, 1961). However, if the damage is chronic, as in cirrhosis, the liver can increase in mass only until the 26th day when cell loss will again balance cell gain, and the situation will stabilize in hyperplasia.

Evidently this type of large-scale regeneration can only occur in such massive tissues as those of the kidney, the liver and the blood. In small tissues and organs the amount of chalone synthesized is probably not sufficient to raise the blood chalone concentration significantly above zero so that compensatory hypertrophy is impossible.

In agreement with this explanation of the mechanism of tissue and organ regeneration, the contrary situation, in which a gross excess of a tissue within the body results in the suppression of mitosis in that tissue, is well known. Such an excess develops during the growth of any tumour, which results in an abnormal rise in the systemic concentration of the chalone of the tissue from which the tumour developed, and leads to the severe depression of mitotic activity in that tissue (9.2.4).

6.2.4 *Periodic tissue hypertrophy*

Some epithelial tissues are peculiar in that they oscillate between periods of hypo- and hyperplasia. The most obvious examples are the accessory and secondary sexual tissues which increase in mass at the onset of a breeding season or at the onset of an oestrous cycle. Such hyperplasia is, of course, initiated by androgenic hormones in the male and by oestrogenic hormones in the female. One of the best understood of these tissues is the lining epithelium of the vagina.

In the absence of any hormonal stimulation this epithelium may be termed hypoplastic: the mitotic activity is negligible, the epithelium is thin (in the mouse c. 3–4 cells thick), and the rate of post-mitotic ageing with keratinization is extremely slow. In the presence of an oestrogenic hormone the epithelium rapidly expands into hyperplasia: the mitotic rate is high, the epithelium is thick (in the mouse c. 12 cells thick), post-mitotic ageing and keratinization is completed in only 30–46 hours (Peckham and Kiekhofer, 1962), and the keratin squames are shed abundantly into the vaginal lumen.

There is an obvious parallel between this response and that seen for instance in damaged epidermis. Basically they must be identical. In the wound response there is both a fall in the chalone concentration and a fall in the responsiveness of the cells to what chalone remains; in the vaginal response the oestrogenic hormone seems to neutralize the chalone itself, or to reduce the cell responsiveness to the chalone, or both. As Jensen (1962) has said, 'an oestrogen may inhibit a growth-restraining factor', which it is now evident may be the chalone molecule itself or its specific receptor molecule.

Since chalone action is tissue-specific the implication is that a mitogenic hormone molecule is tissue-specifically adapted to interact with the chalone mechanism of its target tissue. However, tissues that are not directly associated with reproductive cycles may also show a similar, though much smaller, response to the androgenic and oestrogenic hormones, which indicates that the actions of these hormones may be more widespread than is often supposed. Thus the mitotic rate of the general epidermis increases under both androgenic and oestrogenic stimulus (Bullough and van Oordt, 1950), and the same is true of the sebaceous glands (Ebling, 1957, 1963). This suggests that, during evolution, almost any tissue may become involved in the reproductive process as a target tissue under sex hormone control. Examples are the cock's comb, the frog's thumb pads, and the posterior region of the male stickleback's kidney, which in the breeding season enlarges and produces a secretion that is used in nest building (Bullough, 1946).

Another dramatic example of periodic hypertrophy in an epithelial tissue is provided by the hair bulb. The production of a new hair begins with the sudden onset of extremely high mitotic activity in the basal cells, the extruded post-mitotic cells then keratinizing and dying quickly to form the hair shaft. This activity continues for a fixed period by which time the hair has grown to an appropriate length. The mitotic activity then ceases abruptly and completely, and the hair bulb enters a resting phase. At least in colder countries the production of a new hair coat occurs typically in the spring and in the autumn, but in such animals as rats and mice there is a continuous rhythm, a new phase of hair growth beginning ventrally as the previous phase finishes dorsally (Durward and Rudall, 1958). In these animals follicular activity is synchronized over large areas of the body, but in animals like the guinea-pig and man each hair follicle acts independently to follow its own individual rhythm (Chase, 1954).

Many attempts have been made without success to understand the basic control mechanism of the hair growth cycle. However, it is unlikely that the mitotic control mechanism will prove to be unique, and it follows that a chalone system must somehow be involved. In support of this it is known that damage to a quiescent hair bulb causes an immediate resumption of mitotic activity, which suggests that the previous lack of mitosis was due to high chalone effectiveness. Further evidence comes from the way in which the stress hormones can delay the onset of the next period of hair growth, while conversely adrenalectomy can accelerate it (Bullough, 1965).

Whereas in a normal tissue the rate of chalone production may be constant, it is possible that in the hair root cells it is rhythmic. If this is so, it is also possible that the ultimate control may rest with a group of genes that oscillate between activity and inactivity, in the manner of the oscillating gene control of oestrous cycles (Grüneberg, 1952). Alternatively, the rhythm could have its origin in the cells of the dermal papilla on which the hair bulb is based. In support of this it is evident that the dermis controls the timing of the hair growth cycles. In the different body regions the hairs are of different lengths, which means that their growth periods are of different durations. This is particularly obvious in man in whom the head hair grows to a great length, that is the growth period is long and the rest period is short, while the body hair may be so short and thin that it is almost invisible, which means that the growth period is short and the rest period is long. These regional differences are established in the fetus (5.1.2) but the instructions to the hair root cells may change as when, on a balding head, long hair production is changed to short hair production.

6.3 Tissue organization

As already emphasized, no tissue consists merely of a haphazard mass of cells. Each develops its own specific form of cellular arrangement, or cellular architecture, that is dependent partly on the shape of the connective tissue surface on which the basal epithelium lies, and partly on the route followed by the ageing post-mitotic cells as they move away from this baseline. Beneath a lining epithelium the connective tissue is relatively flat; in a solid epithelial structure the connective tissue has the form of a three-dimensional network.

6.3.1 Epidermis-derived tissues

In the epidermis the intermitotic cells are forced out of the basal cell layer by the pressure of adjacent mitoses, and then, following one of a number of possible patterns of movement, they form a stratified epithelium as they pass to their death distally. The sebaceous gland cells behave in essentially the same way. As cells are forced from their dermal contact they become post-mitotic, increase greatly in size, synthesize sebum, and die and burst as they are pushed out through the gland orifice. Although the gland has the shape of a ball, it too is a stratified epithelium with the mature cells central. Epidermal cells can form a similar structure when, after skin damage, a group of them becomes isolated within the dermis; the peripheral cells are mitotic, and the centre of the ball fills with keratin.

The growth pattern of the hair follicle is similar. The mitotic rate of the basal cells is so high that all the mitoses are angled vertically (4.2.6) with the result that one of each pair of daughter cells is supra-basal, and so enters directly into the rapidly growing column of post-mitotic cells. These cells adhere particularly tightly together as they keratinize to form the hair shaft. Thus a hair is an enormously elongated stratified structure formed by a local accentuation of the normal stratified epidermal structure.

However, the cells of another epidermis-derived tissue, the eccrine sweat gland, behave quite differently. The gland cells themselves show little mitotic activity, but in the cells of the long twisting ducts mitosis is common (Bullough and Deol, 1972). Each duct is lined by a double cell layer: the dermis-adjacent cells of the outer, or basal, layer are mitotic, while the cells of the inner, or supra-basal, layer are post-mitotic and form the duct lining. As usual mitotic pressure forces intermitotic basal cells out into the supra-basal lining epithelium. However, instead of forming a stratified keratinizing epithelium, which would block the duct, the cells, as they insert themselves into the lining epithelium, cause this lining to increase in length and so to move outwards, sliding over the fixed basal cell layer. The ultimate fate of these sliding post-mitotic duct cells has not been determined. Some may age and die while still part of the duct lining, while others may be forced out through the duct orifice to be lost on the skin surface.

6.3.2 *Other epithelial tissues*

A similar sliding movement of the post-mitotic cells is seen, for instance, in the epithelial lining of the duodenal crypts, but in this case, since the epithelium is only one cell thick, the cells slide directly over the surface of the connective tissue. Mitotic activity is confined to the basal crypt region, where the mesenchymal factor concentration may be higher. As usual each cell, when entering the mitotic metaphase, increases greatly in volume, but since the adjacent cells maintain their firm grip on the baseline, the dividing cell is forced to balloon outwards into the crypt lumen while still retaining its grip on the baseline through a thin stalk to an expanded base. Thus a metaphase cell assumes the shape of a wine glass. During the following hour or two the swollen distal cell region, now divided into two, is retracted into the cell layer so providing the tension required to force the sheet of cells to move outwards towards the mouth of the crypt. Beyond a certain point in their sliding journey the stimulus to mitosis is lost (the G_1 chalone action evidently predominates), and the cells become post-mitotic. At the mouth of the crypt they are detached into the intestinal lumen where they die and are digested.

A more complex situation is found among the epithelial cells of such solid organs as the pancreas and the liver. When newly formed in the embryo their structure and pattern of cell behaviour is simple, resembling those of the sebaceous gland. Thus an embryonic pancreas consists at first of a small group of cells surrounded by a simple capsule of connective tissue. The peripheral cells are mitotic while the central cells, separated from the influence of the mesenchymal factor, begin their post-mitotic tissue-specific syntheses (Wessells, 1964a). As development proceeds the epithelial cell mass becomes lobulated and complex, while the connective tissue forms a sponge-like network that permeates the whole structure and contains the blood capillaries. When the pancreas or liver is fully grown the epithelial cells have only a low mitotic rate, but it is probable that such mitoses as do occur are confined to those cells that are in connective tissue contact, as is the rule in the embryo.

6.4 Summary and conclusions

The main single conclusion is that those mechanisms that in epidermis control mitotic activity, initiate tissue function, direct the ageing process in the post-mitotic cells, and determine the cellular architecture, exist in the same general form in all other types of epithelial tissue.

Cellular homeostasis is maintained by a complex control mechanism, the greater part of which is identical in all types of tissue cells; only the G_1 and G_2 chalone molecules that activate this common mechanism are specific to each tissue type. The central feature of all these chalone control mechanisms, the one that ensures continuing cellular homeostasis in all conditions from extreme hypoplasia to extreme hyperplasia, is the close interconnection that persists between the rate of mitosis and the rate of post-mitotic cell ageing. The one rate always varies in step with the other; the rate ratio is constant.

This close similarity between the chalone mechanisms of the epithelial tissues is not surprising: as Rytömaa (1976b) has said, 'it seems unlikely, especially from an evolutionary point of view, that the basic control of cell proliferation would be drastically different in different tissues', and although so little is yet known about it, the same is probably also true of the control of cell movement and thus of tissue organization.

6.4.1 *Cell gain and cell loss*

In all epithelial tissues the rate ratio remains constant over the widest range from pathological hypoplasia to pathological hyperplasia. Indeed a tissue may normally exist at any point in this range. Thus the liver hepatocytes have such a naturally low mitotic rate that this tissue may be termed 'hypoplastic'; its mass cannot shrink further but it can increase into hyperplasia. Conversely, the lining epithelium of the duodenal crypts shows such a high mitotic rate that it lies naturally towards the 'hyperplastic' end of the range.

In all epithelial tissues the replacement of cells lost by damage depends on the fact that, although the increased mitotic rate is immediately offset by a reduced post-mitotic life span, enough new cells are produced within this reduced life span to make good the loss. Such tissue regeneration also involves active cell migration whereby the tissue architecture is, with greater or lesser success, rebuilt. All these processes, the increased mitotic rate, the increased cell ageing, and the active cell migration, are merely accelerations of the normal processes that occur continuously within each epithelial tissue.

The same processes are similarly accelerated for the same reasons during compensatory hypertrophy, the large-scale form of tissue regeneration that is possible only in massive tissues. In this case the responses are seen not only within the damaged tissue but also, due to the fall in the blood chalone concentration, in any similar tissue at a distance. However it is important to note that when, for instance, one kidney is removed from a pregnant animal the regenerative response is seen only in the remaining maternal kidney. It does not occur in the kidneys of the fetus in the uterus (Goss, 1963). It thus seems that maternal chalones do not cross the placenta, and since they are inhibitors of tissue growth this is not surprising.

6.4.2   *Tissue structure*

It is the general rule in epithelial tissues that only connective-tissue-adjacent cells are mitotic. If the influence of the mesenchymal factor is relatively strong its mitosis-promoting action may extend into the lower supra-basal cell layers, as it does in the hair bulb; if the influence is relatively weak even the basal epithelial cells may become post-mitotic, as in the hepatocytes.

Each tissue has its own form of cellular architecture which is determined by the tissue-specific movements of the post-mitotic cells. In general, these cells may either move distally away from the connective tissue baseline or they may slide away sideways. However, the details of their movements are complex, and the way in which tissue-specific cellular architecture is created and maintained remain unknown.

6.4.3   *Tissue mass*

The other aspect of tissue structure is tissue mass. This is determined by a number of factors, the least important being the mitotic rate which merely fixes the position of the tissue on the hypoplasia–hyperplasia scale. An increase or decrease in the mitotic rate has only a relatively small effect on tissue mass.

The second factor, about which little is yet known, is the value of the rate ratio (rate of cell gain : rate of cell ageing = a constant). It is possible that in each epithelial tissue the value of the ratio may be different, and if this is so, then it is clear that the longer the post-mitotic cells take to age and die relative to the mitotic rate (i.e. the smaller the value of the constant) the larger the tissue will be. This is an area for further research.

It is the third factor, the absolute number of basal cells, that is apparently the most important in influencing tissue mass. As already emphasized, the existence of the constant rate ratio implies the existence of a constant number ratio (number of basal mitotic-cycle cells : number of supra-basal post-mitotic cells = a constant). Thus in any epithelial tissue the number of basal connective-tissue-adjacent cells determines the number of supra-basal cells and thus the tissue mass.

The number of basal cells is itself determined by two factors of which the first is the strength of the grip that holds the basal cells to the connective tissue surface. The stronger the grip, the more difficult it is to displace the cells, and the more crowded and columnar they become. It is important to re-emphasize that, although it is the mitotic activity that produces the tension that crowds the cells, such crowding develops irrespective of the mitotic rate.

The second and more important factor determining the number of basal cells is the area of the connective tissue surface on which they lie. Thus the epidermis, although very thin, is a massive tissue because of its huge baseline area, which may be further increased, as in man, by baseline folding. Equally great baseline areas are found in such organs as the liver in which the connective tissue forms an extensive three-dimensional network. The enormous number of basal hepatocytes is then matched by an even larger number of supra-basal cells, and the liver is consequently a massive organ.

6.4.4 *The structure of a chalone mechanism*

One point of particular importance that emerges from this survey is that in any epithelial tissue the proper working of the chalone mechanism depends on the physical segregation of the two types of differentiated cells: the mitotic-cycle cells form one compartment that lies close to the connective tissue while the functional ageing cells form another compartment that lies away from the connective tissue (or, in some cases, away from special regions of the connective tissue). Without this physical separation, created by the opposing concentration gradients of the mesenchymal factor and the G_1 chalone, the mechanism could not work. Thus in the connective tissues, where for obvious reasons this physical separation is impossible, a simple chalone mechanism is not adequate to maintain cellular homeostasis (see Chapter 8).

Chapter 7

Muscles and nerves

7.1   *The muscles*   101
      7.1.1   *Mitotic smooth muscles*   101
      7.1.2   *Non-mitotic muscles*   101
7.2   *The nerves*   103
      7.2.1   *Non-mitotic neurones*   103
7.3   *Summary and conclusions*   103
      7.3.1   *The problem of ageing*   104

Although all the epithelial tissues replace their cells and maintain their mass by means of the same mechanism, or mechanisms, with only the chalone molecules being tissue-specific, there are two other groups of tissues that show marked variations on this basic theme. These are, first, the non-mitotic muscles and the non-mitotic neurones of the central nervous system, and second, the various connective tissues (Chapter 8). Altogether these tissues form the bulk of the body mass.

Theoretically it is possible to envisage an adult animal, created by mitosis and tissue-differentiation, in which the constituent cells are never replaced but remain in a functional state throughout the whole of life. This situation is probably to be found in the tissues of adult insects, which usually have short lives but which may sometimes survive for a year or more in the adult state; it certainly exists in most of the muscles and nerves of adult mammals, although in both these tissues the molecular components of the cells are constantly being replaced.

7.1 The muscles

Both in their structure and in their methods of homeostatic control the muscles fall into two sharply distinct groups: the smooth muscles composed of mononucleate cells which sometimes undergo mitosis, and the skeletal or striped muscles composed of cells which are fused together in groups to form multinucleate fibres in which nuclear division does not occur. The cardiac muscle fibres are also multinucleate and non-mitotic.

7.1.1 *Mitotic smooth muscles*

In an adult mammal mitotic activity in smooth muscle cells, where it occurs at all, is uncommon. However, it is clear that these muscles do have a mitotic potential. This is seen, for instance, in the alimentary canal after wounding (McMinn, 1969), in arterial smooth muscle during atherogenesis (Florentin *et al*, 1973), and in uterine smooth muscle following hormonal stimulation (Bullough, 1946). Mitoses also develop abundantly when, for instance, vascular smooth muscle cells are maintained *in vitro* (Gospodarowicz *et al*, 1981). In all these respects the responses of the smooth muscle cells parallel those of a typical, or a hormone-dependent, epithelium.

Even more significantly it has been found that partially purified extracts of arterial wall contain a mitotic inhibitor that acts specifically on arterial smooth muscle cells made hyperplastic by excess cholesterol (Florentin *et al*, 1973; Thomas *et al*, 1976). This inhibitor has the characteristics of a smooth muscle chalone.

Thus the evidence suggests that smooth muscle cells are subject to a normal mechanism of cellular homeostasis. It appears that the actions of their tissue-specific chalones must be relatively powerful so that the low rate of mitosis is matched, as usual, by a long post-mitotic and functional life expectancy. However, the fact that uterine smooth muscle responds by active mitosis to hormonal stimulation at oestrus and during pregnancy while other types of smooth muscle do not, suggests that there could be more than one type of smooth muscle chalone system. Possibly the different kinds of smooth muscle, like the different kinds of epithelia, are separately controlled.

7.1.2 *Non-mitotic muscles*

The evidence is that the striped and cardiac muscles show no mitotic activity in an adult mammal. Obviously, during the embryonic and fetal periods, when these muscles are being created by mitosis, a cellular homeostatic mechanism must exist, and the question arises whether any part of this mechanism persists into the adult phase.

During their mitotic phase the muscle cells are mononucleate; it is when these cells become fused together in groups to form elongated multi-nucleate fibres with the typical cross-striations that they lose their mitotic potential. Thereafter nothing has yet been found that will induce further mitosis, and it seems that the nuclei have all entered the A_2 phase of the ageing pathway in which the mitosis genes are finally silenced. Also, since there is little or no old cell death, few if any of the nuclei can pass beyond the functional A_2 phase into the dying or D phase.

From the juvenile phase, when mitotic activity ceases, to the adult phase, the massive growth of both the skeletal and the heart muscles is due to the increase in size of the syncytial muscle fibres, and the same process occurs in the adult when a muscle is exercised. In response to an increased workload there is an increased rate of molecular synthesis which increases the mass of the cell contents. This is molecular hyperplasia, and it indicates the existence of a mechanism of molecular homeostasis. At the beginning of a period of extra exercise the energy output per unit of muscle mass is increased; after muscle hypertrophy the normal ratio of energy output to muscle mass is either regained or approached.

After limited damage, as for instance when a muscle is cut, there is no production of new cells or of new nuclei. The cut is closed by muscle fibres growing in length to extend across the damaged region and to interdigitate with fibres on the other side. After heavy damage, as when a substantial part of the muscle is cut away, there is no regeneration. Thus, in the skin, when part of the subdermal panniculus carnosus (a thin sheet of striped muscle) is cut away it never regrows (Bullough and Laurence, 1960b).

Clearly the striped and cardiac muscles represent an extreme variation on the normal theme of tissue cell homeostasis. Zero cell gain appears to be matched by zero cell loss, and it is possible that in such a situation no chalone mechanism is needed. However, the functional tissue genes are obviously working actively under the influence of the molecular homeostatic mechanism, and it is possible that a G_1 chalone may be present to maintain this nuclear activity. Experimentally this possibility could easily be checked.

There also remains the open question as to whether the post-mitotic ageing process has ceased altogether or whether it has merely been slowed to an extreme. In the former case, if there is such a thing as an 'ageing operon', this could have been closed down completely, like the mitosis operon; in the latter case, if the ageing process still continues, albeit at an extremely slow rate, this part of the chalone control mechanism may also still operate (see 7.3.1).

7.2 The nerves

The same non-mitotic nuclear condition is typical, for instance, of the neurones of the central nervous system. The whole stock of cells is built up by mitosis in the embryo and fetus, when some form of mitotic control and of cellular homeostasis must exist. The subsequent growth in bulk of the nervous system through the juvenile to the adult phase is due to the increase in cell size combined with the increase in axon length. Indeed throughout life the rate of synthesis of new macromolecules within the cells remains high. Newly-formed cytoplasm is pushed out along the axon at a speed of about 1 mm per day, the older cytoplasm being catabolized at the axon end.

7.2.1 *Non-mitotic neurones*

It is well known that after any form of damage to the nerve ganglia the lost neurones can never be replaced. Nerve regeneration is limited to the regrowth from the proximal stump of any axon that has been severed, while the separated distal region of that axon degenerates. This process of axon regeneration is evidently based on the normal process of new cytoplasm synthesis.

As with the striped and cardiac muscles, all the non-mitotic neurones must lie on the ageing pathway, and since the mitosis operon is firmly closed down, the cells must be considered to be in the mature functional A_2 phase. The same questions then arise as with the striped muscles: whether the continuing presence of a G_1 chalone is essential to maintain the genetic activity on which the synthesis of new cytoplasm depends; whether the process of post-mitotic ageing has completely stopped, which might imply that the whole cellular homeostatic mechanism has been disconnected, or whether cell ageing is still continuing at an extremely slow speed, which would mean that this part of the chalone control mechanism is still operating (see 7.3.1).

7.3 Summary and conclusions

Those tissues which in the adult mammal have lost the ability to replace their cells, even in an emergency, evidently present only a simple variation on the common theme of cellular homeostasis. In the young animal, when these tissues are still mitotic, their cells must obviously be subject to control by what are probably typical chalone mechanisms, and although in the adult the mitosis genes in the functional cells are firmly closed, there remains the possibility that some part of this chalone mechanism still remains.

7.3.1   *The problem of ageing*

A chalone mechanism has at least two closely related functions: the control of mitosis and the control of post-mitotic ageing. The possibility exists that in these non-mitotic tissues the mechanism of cell ageing is still operating and therefore, from what has been said earlier, that a G_2 chalone may be present and active.

Unfortunately, there is no agreement as to whether cells are or are not lost from the non-mitotic tissues in the course of a long life. Evidence for cell loss includes that of Johnson and Erner (1972), who estimated that, while the brain of a young mouse contains some 5.5 million neurones, after the age of 2 years this figure falls so sharply that in the oldest mice examined there were only about 2 million neurones remaining. Earlier, Curtis (1963) had similarly claimed that, as a rough estimate, the adult human brain loses some 10 000 neurones per day, which is less serious than it may sound because of the enormous number of neurones (perhaps 100 thousand million) that are originally present. It is certainly obvious that in an old mammal there may be a considerable reduction in the size of the brain and in the mass of the skeletal and cardiac muscles, and it has been claimed that this is the crucial factor that limits the life span of the animal (Bullough, 1967, 1973; Johnson and Erner, 1972).

However, directly contrary evidence has been given by Franks *et al* (1974), whose results indicated that in the old mice they studied there was no reduction in the wet or dry weight of the brain or in its protein or DNA content. The final answer remains in doubt but there does also exist some interesting indirect evidence.

If the length of life of a mammal is determined by the length of time that the brain and the skeletal and cardiac muscles continue to function effectively, then it should be possible to lengthen the life span by any technique that will delay the ageing of the cells within these tissues. In an epithelial tissue the one well-established technique for delaying cell ageing and inhibiting cell death is to increase chronically the stress hormone content of the body. In relation to the ageing of whole animals this type of experiment has been repeatedly performed and the results are not in doubt.

In long series of experiments by many investigators it has been found that when mice or rats are chronically stressed by partial starvation the mass of their adrenal glands is greatly increased and they enjoy a life span that is up to twice as long as that of their well-fed litter mates (see Tannenbaum and Silverstone, 1957). It is also significant that these stressed animals do not simply survive longer in old age; the signs of old age are delayed and the animals remain younger longer. In this state they remain resistant to disease and the onset of all forms of cancer is postponed (9.1.3).

At first this dramatic result was ascribed to the effects of the diet *per se*, but it is now clear that it was in fact due to the imposed stress. A similar result can be obtained simply by administering a glucocorticoid hormone in the drinking water throughout the whole of adult life (Bellamy, 1968). However, it is also clear that the level of stress imposed must be only moderate; if it is too great it brings its own problems which can prove fatal.

These experiments are suggestive but not conclusive. However, they are in agreement with a theory that these critically important muscular and nervous tissues, by becoming non-mitotic, have created a situation whereby their slow tissue-specific rates of cell loss have set a species-specific limit to the life span of each mammal. The loss of cells from these non-mitotic tissues would at first cause little reduction in efficiency because of the great reserve metabolic potential provided by the molecular homeostatic mechanism, but beyond a critical age, when this reserve had been significantly eroded, the weakening of the skeletal and cardiac muscles and of the nervous responses would diminish the efficiency of most if not all the systems of the body. In the wild, death is certain when an animal is no longer able to meet the normal daily or seasonal crises; in man and his domestic animals, old age is an extreme and artificial condition that is never reached in nature (see 10.3.3).

If this theory is correct its main value would lie in the opening it would provide for an experimental analysis of ageing, which still remains the most obscure of biological problems.

Chapter 8

The connective tissues

8.1		*The structural connective tissues*	107
	8.1.1	*Dermal cell mitosis* in vitro	108
	8.1.2	*Dermal response to wounding*	109
	8.1.3	*The nature of the mitotic cells*	110
8.2		*The blood cells*	111
	8.2.1	*The origins of the blood cells*	112
	8.2.2	*The granulocyte system*	114
	8.2.3	*The granulocyte chalone*	115
	8.2.4	*The erythrocyte system*	116
	8.2.5	*The macrophage and megakaryocyte systems*	117
	8.2.6	*The lymphocyte system*	118
	8.2.7	*Tissue transplantation and rejection*	120
8.3		*Summary and conclusions*	120
	8.3.1	*The connective tissue chalone mechanisms*	121
	8.3.2	*Cellular homeostasis in connective tissues*	121
	8.3.3	*The poietins and antigens*	122
	8.3.4	*Chalone tissue-specificity in blood tissues*	123

In the epithelial tissues the proper functioning of the chalone control mechanism depends on the physical separation of the mitotic-cycle cells from the post-mitotic cells. In the connective tissues, even if the cells are able to respond to the mitotic stimulus of their own mesenchymal factor, it is obvious that the mitotic-cycle cells, being entirely embedded in connective tissue, cannot form a physically separate cell population. It is this difficulty that has led to a second variation on the theme of the chalone control mechanism.

The various types of connective tissues originate from the embryonic mesenchyme, which consists of multipotential stem cells, and in the adult they fall into two main groups, the structural connective tissues and the various blood cell tissues. The lymphocyte system also belongs with the blood cell tissues although its embryonic origins are more complex.

Of all the connective tissues, the best studied from the present point of view have been the dermis, in which mitosis is normally rare, and the various types of blood cell populations, in which there is a high rate of new cell recruitment and of old cell loss.

8.1 The structural connective tissues

Although there have been some preliminary studies of cell replacement and cell behaviour in cartilage and bone, attention has been mainly concentrated on the reticular type of connective tissue as exemplified by the dermis. Any study of reticular connective tissue faces two major problems (see Glücksmann, 1964; McMinn, 1969; Pritchard, 1974). First, there is the confusion caused, for instance in the dermis, by the wide variety of cell types that are present: besides the connective tissue cells themselves, there is a large migratory cell population which includes lymphocytes, granulocytes, monocytes, mast cells, plasma cells, macrophages etc., and a large static cell population which includes capillary endothelial cells and nervous and sensory cells. An additional difficulty is that when a cell is passing through mitosis its identity is often obscured.

Second, there is the problem of nomenclature: the connective tissue cells are commonly called fibroblasts, a name which besides implying an ability to produce collagen fibres also implies an embryonic-type ability to multiply and to differentiate further. It is, however, more than doubtful whether the mature dermal cells are even capable of mitosis, and it may be more realistic to distinguish between fibroblasts which can divide and fibrocytes which cannot. However, yet another type of cell seems also to be involved. There is strong evidence that throughout the body there exists a population of multipotential connective tissue stem cells, or embryonic-type mesenchymal cells, which are mitotic and which by a terminal differentiation can give rise to any of the various cell types of the structural connective tissues. They may differentiate into fibroblasts and so into connective-tissue-synthesizing fibrocytes, into chondroblasts and so into cartilage-synthesizing chondrocytes, or into osteoblasts and so into bone-synthesizing osteocytes. The final explanation of the nature of all the various types of connective tissue cells has yet to be given.

In studying the method of replacement of the dermal cells it is necessary, because of the normal rarity of mitosis, to concentrate on the mitotic activity that occurs in abnormal situations. The first of these is when cells from the dermis, or other similar reticular connective tissues, are maintained *in vitro*, while the second develops after wounding when certain dermal cells respond by mitosis.

8.1.1 *Dermal cell mitosis* in vitro

Cells derived from fetal or adult dermis multiply actively when kept in suitable *in vitro* conditions. From what has been said above the precise nature of these cells remains in doubt: they could be typical fibrocytes that have been induced to revert to mitosis or they could be some special type of mitotic cell, whether fibroblast or mesenchymal cell, which is normally present in only small numbers but which *in vitro* becomes dominant. The only certainties are that the cultured cells were originally associated with the dermal connective tissue and that in the literature they are called fibroblasts.

When cultured *in vitro* such fibroblasts, whether from established cell lines or from adult dermis, require the presence of serum before they will enter mitosis. It was using such a system that Houck (see Houck and Daugherty, 1974) demonstrated that the serum mitogen is a glycoprotein composed of two identical subunits, each of 60 000 daltons. More important he also found that old medium in which fibroblasts have been cultured, or alternatively a water extract of cultured fibroblasts, contains a fibroblast-specific mitotic inhibitor, or fibroblast chalone, and he concluded that this chalone probably acts by competing with the serum factor for the same receptor sites on the cell membrane. The existence of a fibroblast chalone in dermal cells, and in other similar reticular connective tissue cells, has been confirmed by much other work including some in which the inhibitor has been obtained directly from dermal extracts (see Iversen, 1981).

The only fibroblast chalone so far described inhibits the G_1-S transition and is, therefore, a G_1 chalone; no search has yet been made for a G_2 chalone. This G_1 chalone was first thought to be a protein or glycoprotein of at least 40 000 daltons, but as with other G_1 chalones, estimates of its molecular weight have been steadily reduced (Patt and Houck, 1980). The latest information suggests that the fibroblast chalone is a peptide of some

2000 daltons or less. As with other chalones, the high original estimates of the molecular weight were due to the strong tendency of the chalone to aggregate with other larger molecules and especially with proteins and glycoproteins.

Such aggregation may help to anchor the chalone molecules in and around the cells, and Houck has suggested that the fibroblast chalone may normally exist as a cell surface component. Thus an initial conclusion may be that dermal fibroblasts, whatever their precise nature may be, are subject to the usual type of chalone control mechanism with a general mitotic stimulant acting in competition with a tissue-specific G_1 chalone, perhaps at special sites on the outer cell surface.

8.1.2 *Dermal response to wounding*

For the dermal fibroblasts to enter mitosis *in vitro* the concentration of a mitogen, perhaps the mesenchymal factor, must be increased by the inclusion in the medium of a high concentration of serum; conversely, for them to enter mitosis *in vivo* it would be expected that a fall in the G_1 chalone effectiveness would be necessary, and that this would occur after any form of dermal damage. However, the mitotic response of the dermal cells to wounding is more complex than this.

Any skin wound, if deep enough, breaks through at least three distinct connective tissue layers. The first is the thin dermal papillary layer in which connective tissue cells are abundant; the second is the thick dermal reticular layer in which the cells are more widely spaced; and the third and deepest is a layer of fatty connective tissue, the panniculus adiposus. Mitosis is normally rare in both the papillary and reticular layers; in the fatty layer it is evidently absent altogether. After tissue destruction re-generation occurs in the two outer dermal layers but the fatty tissue may not be replaced (Bullough and Laurence, 1960b).

The response to dermal wounding can be summarized as follows (Bullough and Stolze, 1981, unpublished data). On day 1 there is a massive invasion of leukocytes accompanied by oedema. On days 2 and 3 there is a great increase in the number of dermal connective tissue cells in spite of the fact that the rise in the mitotic rate is only slight. On day 4 the mitotic rate suddenly rises to a maximum (more than 100 × normal), a response which *follows* the rise in the dermal cell numbers. The clear implication is that the new dermal cells have come into the tissue from elsewhere, and that only after their arrival do they undergo one or two mitotic divisions. By day 5 the mitotic rate is falling rapidly towards its normal low level.

These observations fit well with the various descriptions that have come from histological studies of the wound healing process (McMinn, 1969; Winter, 1972). In summary, the new dermal cells are first seen in or around the adjacent blood capillaries, which are also regenerating. They then move inwards towards the wound cavity and undergo mitosis before transforming into collagen-synthesizing mature fibrocytes. The conclusion has been that these new cells are derived either from resting fibroblasts lying around the blood capillaries, or, more probably, from undifferentiated mesenchyme cells that may also lie around the blood capillaries or that may arrive via the blood stream from some distant cell depot. Certainly from all the evidence it seems that the dermal fibrocytes, that were part of the papillary and reticular layers before the wounding occurred, do not themselves undergo mitosis or otherwise contribute to the healing process.

8.1.3   *The nature of the mitotic cells*

The difficulty of distinguishing and defining the different groups of dermal cells has been emphasized, but obviously a full understanding of this problem is necessary for any final analysis of the cellular homeostatic mechanism. The only cells that can be identified with any certainty are the mature and functional fibrocytes, which are probably in the A_2 phase of the ageing pathway.

It is the nature of the invading cells that undergo mitosis that poses the problem. All the indications are that these cells move in from cell depots that adjoin the local capillaries or that lie at a distance somewhere in the body, and that they are multipotential stem cells, which means that they are embryo-type mesenchymal cells in which the final step in differentiation has not yet been taken. It has been suggested that this final step is taken in response to the type of connective tissue environment in which they come to lie (Bullough, 1975). Those entering the dermis differentiate into dermal connective tissue cells, which then undergo at least one mitosis before being inhibited by the dermal chalone (the so-called fibroblast chalone) and maturing into functional fibrocytes. Such cells could not, of course, ever regain the previous multipotential state.

Support for this theory comes from the remarkable observation that, when a polyhydroxyethylmethacrylate sponge is inserted within the dermis, the cells that arrive during the consequent dermal healing process are converted not into fibrocytes but into osteocytes, which then lay down bone within the sponge (Winter and Simpson, 1969). Evidently the abnormal environment induces an abnormal terminal differentiation, which clearly shows the multipotential capability of the incoming cells.

In this connection Pritchard (1974), describing the development of cartilage and of bone, has also concluded that the incoming mesenchyme cells, or stem cells, are at least bipotential in that they may be converted into either chondrocytes or osteocytes. He notes that during bone growth, the stem cells, from which the osteocytes are recruited, lie clustered around the blood capillaries exactly as they seem to do in the dermis.

Taking all the evidence together the tentative conclusions must be that in a structural connective tissue the supply of new cells is provided by a population of multipotential stem cells; that on entering a connective tissue these cells undergo a terminal differentiation into the appropriate type of connective tissue cell; and that only after this final step do they undergo mitosis and come under the control of the specific connective tissue chalone mechanism. The stem cell population, wherever it is located, must also sustain itself by mitotic activity that is itself controlled. All these interlocked processes occur slowly to support normal connective tissue cell replacement, and quickly to support regeneration after connective tissue damage.

It is also possible that all the structural connective tissues of the body are dependent on a common stock of stem cells, and even that these cells are in constant movement around the body.

These various theories lack proof which may not be easy to obtain. However, a precisely similar control system is known to maintain the various connective tissues of the blood; one or more multipotential stem cell populations providing, by terminal differentiation, the various chalone-controlled blood cell types. It has even been suggested that the stem cells of the structural connective tissues may be related to, or identical with, the stem cells of the blood tissues (Halpap and Cremer, 1972).

8.2 The blood cells

The cells of the haemopoietic system are centred in the bone marrow from which, when mature, they are released into the blood (Lajtha, 1973); the cells of the lymphocyte system are centred in a variety of depots (bone marrow, thymus, spleen, lymph nodes) from which they pass into the lymph and blood. The lymphocytes and their chalone systems are considered in 8.2.6.

8.2.1 *The origins of the blood cells*

Within the bone marrow embryonic multipotential stem cells persist throughout life, and from them constant supplies of granulocytes, erythrocytes, megakaryocytes and macrophages are formed by typical terminal differentiations. These various kinds of terminal blood tissues are peculiar in that the mitotic activity of their cells, although considerable, is not adequate to maintain the tissue mass. Without constant recruitment from the stem cell population these tissues would vanish. Clearly the haemopoietic system is complex, and it is controlled and maintained in balance by a considerable number of regulatory signals. Essentially this regulation is exercised at four levels (see Rytömaa, 1976a).

First, the population of multipotential stem cells is maintained by a mechanism which ensures that the rate of cell gain by mitosis keeps pace with the rate of cell loss by differentiation into the various terminal blood tissues. This is evidently achieved by a typical chalone mechanism, that is through the action of a specific stem cell chalone (Iversen, 1981). This chalone was first reported to have a molecular weight of some 50 000 daltons or more, but more recent evidence suggests, as usual, that this high estimate was due to molecular aggregation, and that the true figure may be < 1000 daltons. The stem cell chalone does not inhibit mitosis in the terminal blood tissues, and the various chalones of these tissues do not inhibit mitosis in the stem cells.

Second, there is the terminal differentiation of the stem cells into the so-called 'committed cells', that is into cells that are committed to form a particular end tissue, for instance granulocyte cells, and so to synthesize their own tissue-specific chalone. Presumably this step is initiated, as in any newly-forming embryonic tissue, by the action of a specific inducer molecule. The possibility has been suggested that in the adult mammal this inducer may be synthesized by the already committed cells.

Third, there is the mitotic control exercised over the newly formed committed cells. These cells undergo a series of mitotic divisions which amplify the cell population until this activity is suppressed by their tissue-specific chalone. Thus progranulocyte multiplication is limited by the granulocyte chalone, pro-erythrocyte multiplication by the erythrocyte chalone, and so on.

Fourth, the committed cells are accelerated to maturity under the stimulus of specific humoral factors, called poietins. Thus haemoglobin is actively synthesized in the erythrocytes under the stimulus of erythropoietin, a glycoprotein of 46 000 daltons, which is itself synthesized, especially in the kidneys, in response to oxygen lack. In the granulocyte system the granulopoietin is probably the so-called 'colony stimulating factor', which is necessary for granulocyte production when bone marrow cells are cultivated *in vitro*. A thrombopoietin has also been described, and this may be related to megakaryocyte maturation, and therefore to platelet production.

Taking the erythrocytes as an example, the system then operates as follows. A need for oxygen causes the release of erythropoietin into the blood. This passes to the bone marrow, and stimulates the maturation of committed pro-erythrocytes into functional erythrocytes, which then enter the blood to carry more oxygen. This reduces the number of committed pro-erythrocytes in the bone marrow, and there are two consequences. The first is that new committed cells are formed by terminal differentiation from the stem cell population; the second is that the temporary local shortage of erythrocyte chalone, caused by the rapid withdrawal of large numbers of erythrocytes from the bone marrow, allows the newly-formed committed cells to undergo more than the usual number of mitoses before being inhibited. The population of committed cells then rapidly recovers, while that of the stem cells is reduced. There is then a fall in the local concentration of the stem cell chalone and the stem cells multiply to make good their loss.

Thus the control of cellular homeostasis in the end tissues of the haemopoietic system differs from that of an epithelial tissue in two main respects. The first is that each blood tissue is not self-supporting but is composed of cells that are in transit; only the basic stem cell population is self-renewing by means of a typical epithelial type of chalone mechanism. The second is that the blood tissues respond to extra control factors, the poietins, whereby their rate of cellular maturation, and thus their level of function, is adjusted to physiological needs; oxygen lack increases the number of circulating erythrocytes while infection increases the number of circulating granulocytes.

It must be added that any review of the extensive literature on the haemopoietic system will reveal a picture that is considerably more complex than that outlined above. Many other factors have been described as being involved in the various control mechanisms, but as Rytömaa (1976b) has emphasized, most of these are probably influencing factors rather than controlling factors. They equate with the long list of 'growth controlling substances' already described and dismissed above (1.5).

8.2.2 *The granulocyte system*

There remains to be considered in greater detail the method of mitotic control in the committed cells of the bone marrow. As in the other blood tissues, the immature granulocytes pass through a mitotic phase before they mature, and the number of their mitotic cycles depends inversely on the density of the mature granulocyte cell population within the bone marrow.

The main function of the granulocytes is to engulf and destroy foreign bodies, whether living or dead, and for this purpose the cells are distributed throughout the whole body, in the blood and in the tissues. New granulocytes enter the circulation at a relatively low level at all times; in man the normal number entering per day is estimated at about 1×10^{11} (Rytömaa, 1976b). The rate of granulocyte release from the bone marrow is directly related to the concentration of a blood-borne leukocytosis-inducing factor, which increases during an infection or when tissues are damaged or inflamed.

The reserve of mature granulocytes within the bone marrow is then depleted, and the number of mitotic divisions undergone by the newly-formed immature granulocytes is increased to make good the deficiency. It is this response that is controlled by a chalone mechanism.

Due mainly to the work of Rytömaa and his co-workers (see Rytömaa, 1976a, 1978), the granulocyte chalone system is now the best known of all such systems. Originally it was thought that the systemic chalone concentration, maintained by the enormous number of circulating granulocytes, was the critical factor in the control, by a feedback loop, of the mitotic activity of the immature granulocytes in the bone marrow. However, it is now believed that the systemic chalone concentration is relatively unimportant, except in pathological conditions (9.2.4). This is because, although the number of circulating granulocytes is enormous, the number of reserve granulocytes packed within the narrow spaces of the bone marrow is more than thirty times as great. Thus although the systemic chalone concentration is considerable, the chalone concentration within the bone marrow must be much higher, and the message coming from the circulating granulocytes must be overwhelmed by that coming from the reserve granulocytes.

Therefore, it is the reduction in the bone marrow granulocyte population that leads, through a reduced local chalone concentration, to the increased number of mitotic divisions seen in the immature cells, each extra mitotic division doubling the new cell population. Rytömaa (1976b) has emphasized that without this mechanism the rate of increase in the production of new granulocytes at the onset of a serious infection would depend mainly on the rate of their recruitment from the stem cells, which would involve a delay of several days and would, therefore, be dangerously slow.

However, the granulocyte chalone mechanism has two peculiar features. The first is that, despite extensive studies, no sign has been found of a G_2 chalone, and with this goes the fact that adrenalin, which is believed to strengthen G_2 chalone action, has no effect on the rate of granulocyte production.

The second peculiarity, which may be related to the first, is that the mature granulocyte seems to have no fixed post-mitotic life span. Its time spent in the bone marrow depends on the demand for its release, and its life spent in the circulation depends on how long it takes for it to meet and engulf a foreign body. When this has been done a granulocyte commits suicide: it migrates to one of the intestinal villi and so passes into the intestinal lumen, where it is digested. Whether in the absence of release into the circulation, or in the absence of any foreign body, it would prove to have a particular life span is unknown and may be an impossible question to answer. However, the absence of a G_2 chalone and the evident absence of a tissue-specific post-mitotic cell life span may be interrelated (2.4.3).

8.2.3   The granulocyte chalone

The granulocyte G_1 chalone, discovered by Rytömaa and Kiviniemi (1968a), has been extensively studied (see Rytömaa, 1978; Iversen, 1981). It can be obtained from serum, by washing or homogenizing granulocytes, or from leukaemic cells. It has been found to be strictly tissue-specific in inhibiting the onset of DNA duplication in immature granulocytes whether *in vitro* or *in vivo* or, most revealingly, in diffusion chambers implanted in the body cavity (Vilpo, 1979). Much basic information has also come from studies of leukaemic granulocytes in which it also inhibits mitosis (9.2.1).

The latest information indicates that this chalone could be an acidic peptide (Paukovits and Hinterberger, 1978; Balazs *et al*, 1980). According to Paukovits and Laerum (1982), 'its functional groups have been identified qualitatively and quantitatively as being a single primary amino-group which is not the N-terminal group, a single thiol group and three carboxyl groups', which is 'corroborated by the identification of

Asp, Glu, Cys and Lys in dansylated hydrolysates'. Several possible candidate peptides have been synthesized of which pGlu-Glu-Asp-Cys-Lys-OH showed similar activities to the granulocyte chalone, especially *in vitro*. This has been independently confirmed by Foa (1982, personal communication), but it is probable that this substance is not the actual granulocyte chalone.

Experiments have suggested that this chalone cleaves to specific receptor sites on the cell surface (Paukovits and Paukovits, 1975) and that the inhibition thus induced can be reduced simply by washing the chalone molecules from these sites. The granulocyte chalone is unusual in that it does not cleave to macromolecules such as glycoproteins, as do most other known G_1 chalones. It is active at a high dilution which approximates to that of a hormone in its target tissue; it inhibits progranulocyte mitosis at concentrations as low as 1 pg/ml *in vitro* and 1 μg/kg *in vivo* (Rytömaa and Toivonen, 1979).

This emphasizes the main problem that has bedevilled attempts to purify this and other chalones from tissue extracts, and so to analyse their structure. In the case of the granulocyte chalone, on the assumption that the molecular weight is about 1000 daltons, it has been calculated that to obtain 100 μg it would be necessary to start with 1000 kg of fresh granulocytes (Maurer, 1975).

8.2.4 *The erythrocyte system*

The cellular homeostasis of the red blood cells is achieved in essentially the same way as that of the granulocytes. The immature cells in the bone marrow undergo a sequence of mitoses to increase the new cell population before beginning the synthesis of haemoglobin. This synthesis is stimulated by erythropoïetin, perhaps by an amplification of the stimulus to maturity exercised by the G_1 chalone. The erythrocytic G_1 chalone, discovered by Kivilaakso and Rytömaa (1971; and see Iversen, 1981), also limits the number of mitoses undergone by each immature erythrocyte.

Like the granulocyte system, the erythrocyte system contains no G_2 chalone. In relation to this it is also insensitive to the usual antimitotic influence of adrenalin, and the rate of ageing of the post-mitotic cells seems not to vary with the mitotic rate. The length of life of the immature and maturing erythrocytes in the bone marrow is determined by the demand for new erythrocytes in the blood; when in the circulation all the erythrocytes lack nuclei, and are therefore in the dying phase, which has a fixed duration.

However, the erythrocyte system shows one marked difference from the granulocyte system: the proportions of the cell population in the bone marrow and in the circulation are reversed. There are some thirty times as many circulating erythrocytes as there are immature and mature erythrocytes in the bone marrow. Thus in this tissue the main source of erythrocytic chalone is the circulating erythrocyte population, and it is evidently the systemic chalone concentration that is dominant in controlling, by a feedback loop, the mitotic activity of the immature erythrocytes in the bone marrow.

The daily fine adjustment of the numbers of erythrocytes in the blood is therefore achieved by a double mechanism: the number of newly-formed erythrocytes is inversely proportional to the concentration of G_1 chalone, that is to the number of circulating erythrocytes, while the number of maturing erythrocytes is directly proportional to the concentration of erythropoietin, that is to the oxygen demand of the tissues. In the crisis of a sudden acute haemorrhage the speed at which new erythrocytes are produced is greatly increased by the extra number of mitoses undergone by the immature erythrocytes in response to the fall in the chalone concentration, each extra mitosis resulting in a doubling of the cell population.

The erythrocyte chalone has been partially purified, which has been a slightly simpler task because of the relative ease with which large quantities of red blood cells can be collected. Its assay *in vivo* is also easier since the nucleated and potentially mitotic circulating erythrocytes of the chick embryo can be used (Perrins and Jones, 1981). Like the granulocyte chalone, it appears to be a simple peptide with a molecular weight of some 600 daltons, and it too does not combine with larger molecules. It is highly active at low concentrations; one injection into a mouse of only 10 μg of an impure preparation produces a maximum mitotic inhibition.

8.2.5 *The macrophage and megakaryocyte systems*

The other cell types that derive from the bone marrow stem cells have not yet been adequately studied from the point of view of their cellular homeostatic mechanisms. However, there is evidence that during macrophage production, which increases in response to inflammation, the mitotic activity of the immature cells is controlled by a specific macrophage chalone (Laerum and Maurer, 1973; see Iversen, 1981). Preliminary attempts at purification have indicated that this chalone has a molecular weight of less than 10 000 daltons. It has also been confirmed that macrophage mitotic activity is not inhibited by granulocyte chalone.

Still less is known about the megakaryocytes from which the platelets are produced in the bone marrow, although there is some slight evidence that here too a specific chalone system may exist (Iversen, 1981).

8.2.6 *The lymphocyte system*

Although the lymphocyte system is related embryologically to the haemopoietic cell system, in an adult mammal such as man it operates under a separate series of highly complex controls. The lymphocytes themselves are multipotential stem cells and are of two kinds: the T cells derived from the thymus, and the B cells originating in the bone marrow and in various lymph node centres. In all these various regions the mitotic rate is high, which indicates that the post-mitotic daughter cells circulating throughout the body must have a short life span. Many of these post-mitotic cells die and are absorbed by macrophages; others, in the presence of invading foreign material, whether living or dead, undergo a terminal differentiation into plasma cells. Such foreign material is antigenic, and in Medawar's words (1963), 'the antigen, like the (embryonic) inducer, is an agent that commits a cell to a certain pathway of differentiation – to one path among several that it might have taken'. In other words, each type of antigen induces some of the multipotential stem cells to undergo a terminal differentiation into a specific type of plasma cell tissue.

When newly formed the cells of such a tissue multiply by mitosis and so augment the population; when post-mitotic and mature those derived from T cells combine with macrophages to mount a specific attack on the antigenic material, while those derived from B cells synthesize an antibody which, spreading through the circulation, has a high specificity against the antigen.

Thus in its outline the lymphocyte system is closely similar to the haemopoietic system. It is based on a population of multipotential stem cells which are self-maintaining and which, by embryo-type induction, differentiate to produce a range of end-tissues. These, although they undergo some mitotic activity, are not self-maintaining. In this system there are two main levels at which specific chalone mechanisms could operate: in the maintenance of cellular homeostasis in the two types of stem cells, and in the control of the multiplication of each of the many types of plasma cells.

Regarding the plasma cells little is known except that mitotic control mechanisms must exist, and that each must be specific to its own plasma cell tissue. Adding to this the fact that the plasma cell life history is so closely similar to that of the erythrocytes and granulocytes, beginning with a mitotic interlude to augment the cell numbers and ending with post-mitotic maturity, it is reasonable to suppose that a chalone control mechanism, specific to each type of plasma cell, must exist.

Regarding the stem cells, on the other hand, a great deal is now known (see Rytömaa, 1978; Iversen, 1981). These cells have been extensively studied, especially in relation to the chalone inhibition of mitotic activity in the lymph node centres, in lymphocytic leukaemic cells, and in cultures of lymphocytes that have been stimulated to enlarge and divide by the presence either of foreign lymphocytes or of phytohaemagglutinin. This last substance, also called PHA for short, is obtained from a plant extract, and its mode of action in promoting lymphocyte mitosis remains mysterious.

The earliest experiments on lymphocyte mitotic control showed, first, that extracts of pig lymph nodes inhibit the onset of DNA duplication in PHA-stimulated human lymphocytes (Moorhead et al, 1969), and second, that extracts of sheep lymphocytes inhibit the onset of mitosis in mouse lymphocytic leukaemic cells (Bullough and Laurence, 1970b). These antimitotic actions were evidently tissue-specific, and, therefore, the lymphocyte control system must include both a G_1 and G_2 chalone. However, all the many experiments that have subsequently been carried out have concentrated only on the G_1 chalone (Houck, 1978; Iversen, 1981).

Active extracts of the G_1 chalone have been prepared from lymphocytes, lymph nodes, thymus, spleen and malignant lymphomas, and in all cases they have been found to be strictly lymphocyte-specific in their antimitotic action. In addition, there is some indication that there may be two distinct, or partially distinct, chalones, the one controlling the T cells and the other the B cells (Houck, 1978). Thus thymus extracts have been described as acting only against T cells, although extracts of B cells have not given such clear-cut results. This important question remains to be answered.

The repeated attempts that have been made to purify the lymphocyte chalone, or chalones, have resulted in the original molecular weight estimates of up to 50 000 daltons being reduced to the present estimate of about 600 daltons (Patt and Houck, 1980). They have also resulted in the original conception of a relatively large glycoprotein molecule being reduced, as with the other G_1 chalones that have been studied, to a belief that the lymphocyte G_1 chalone may be a small polypeptide. This either normally aggregates with other larger molecules or becomes aggregated during the extraction processes.

8.2.7 *Tissue transplantation and rejection*

From these chalone studies one point of considerable potential interest derives from the close involvement of the lymphocyte system in the rejection of transplanted tissues and organs. A clinical answer to this problem of rejection is to administer a glucocorticoid hormone, although this has many undesirable side effects throughout the whole body. Evidently this hormone inhibits lymphocyte action through its prevention of lymphocyte and plasma cell multiplication, and the belief must now be that this is achieved primarily through the strengthening of the anti-mitotic actions of the lymphocyte and plasma cell G_1 chalones. Certainly hydrocortisone is known to strengthen the action of the lymphocyte G_1 chalone (Balazs, 1979).

It follows that a better way to inhibit the rejection of transplanted tissues and organs, and also perhaps to counter various auto-allergic conditions, would be to use the lymphocyte G_1 chalone itself since this would act on the lymphocyte system alone and leave all the other tissues of the body unaffected. Many experiments have already shown that lymphocyte chalone extracts are able to inhibit allograft and xenograft rejections as well as graft-versus-host responses (Mathé, 1972; Houck and Patt, 1981). Indeed, the prevention of graft rejection, as for instance of rabbit skin grafted on to mice, has been used as the basis for a lymphocyte chalone assay system.

This is one of several reasons why the chemical characterization of the lymphocyte chalone, and its subsequent synthesis in adequate quantities for clinical testing, is urgently needed.

8.3 **Summary and conclusions**

The main conclusion regarding the various connective tissues is that, unlike the epithelial tissues, they are not capable of self-maintenance by their own mitotic activity. To maintain cellular homeostasis each of them is dependent on constant cell recruitment, by an embryonic type of terminal differentiation, from a mutipotential stem cell population, which is itself evidently self-supporting in the manner of an epithelial tissue. In man there are probably some three or four different types of stem cells, all originally derived from the embryonic mesenchymal stem cells. One type is spread along the blood vessels, and provides cells for the various structural connective tissues; one lies in the bone marrow, and gives rise to the main blood tissues; and one, or perhaps two, are centred in the various lymphoid depots, and supply the specialized plasma cell clones that counter invasions by foreign organisms or foreign materials.

8.3.1 *The connective tissue chalone mechanisms*

It is probable that each stem cell population is based on connective tissue, and is maintained by a typical chalone mechanism. Cells lost by death or by terminal differentiation are replaced by chalone-controlled mitotic activity; cell gain and cell loss remain in perfect balance.

When a stem cell takes its final step in differentiation this step is, of course, irreversible; a differentiated connective tissue cell can never be converted into yet another type of tissue cell. Also, as this final step is taken, the cell loses its responsiveness to the stem cell chalone and acquires a responsiveness to the chalone system of the end-tissue to which it then belongs.

In the various blood cell tissues the chalone systems are strictly tissue-specific (8.3.4); but whether this is also true of the various structural connective tissues is not yet known. Chalones have been detected not only in the dermis (Hanks, 1978) but also in cartilage (Hardrup *et al*, 1975) and bone (Videman *et al*, 1978); no tests for specificity have yet been made.

8.3.2 *Cellular homeostasis in connective tissues*

The reason why embryonic-type stem cell populations should continue to exist in the adult mammal, and why only the connective tissues should continue to depend on this complex method of cell replacement, has already been suggested.

Essential to a typical chalone homeostatic mechanism are the opposed concentration gradients of the mesenchymal factor and the G_1 chalone, which create the localized communities of mitotic-cycle cells and of mature cells. These gradients are not present in the connective tissues in which, because of the uniform strength of the chalone influence, all the cells are either in, or are rapidly moving towards, the same state of post-mitotic maturity. Only the immature connective tissue cells, when newly-formed from the stem cell population, are transiently capable of mitotic activity during the short interval before they pass into the A_2 phase of the ageing pathway and so lose their mitotic potential.

Just as an epithelial tissue when deprived of its connective tissue base is doomed to disappear, so a connective tissue, which has no such base, cannot by itself survive. Constant cell recruitment from a stem cell population is the only way in which the tissue can be maintained. An inducing agent reacts with a responsive cell, those regions of the genome that are not relevant to the final tissue type are firmly closed down, and those genes on which the final tissue type depends are activated. Some of these activated genes then promote the synthesis of the tissue chalone, while others promote the synthesis of the receptor sites with which the chalone specifically interacts.

The nature of the various inducing agents that act on the connective tissue stem cells is unknown. In the structural connective tissues it seems that some agent is present in each tissue to instruct and convert the incoming cells; in the bone marrow it was once thought that the poietins might be the inducing agents but this now seems unlikely; in the lymphocyte system the inducing agents are evidently the antigens.

8.3.3 *The poietins and antigens*

An additional complication in the cellular homeostatic mechanism of the non-structural connective tissues involves the actions of the poietins on the blood tissues and of the antigens on the lymphocytes. In both these tissue types the end-tissues have not just to maintain a stable cell mass but must also be able to respond rapidly by the creation of extra cells to meet any sudden physiological demand.

As already explained (6.2.3), it has been thought that an increased physiological demand will automatically lead to an increased mitotic rate and so to an increased cell mass, but this can no longer be maintained. To obtain a larger tissue mass an extra mechanism is required, and at least three different answers to the problem can be identified. In skeletal muscles responding to extra exercise the cells increase in mass but not in number; extra amounts of mRNA and of enzyme systems are synthesized, and these occupy extra space. In blood cells, such as granulocytes and erythrocytes, the relevant poietins, formed in larger quantities, accelerate the maturation of the immature cells and so increase the numbers of functional cells. In the lymphocyte–plasma cell system an invading antigen induces the differentiation of a clone of plasma cells that did not previously exist (or reinforces one that did exist) and that is specifically committed to counter that antigen.

8.3.4 *Chalone tissue-specificity in blood tissues*

In the definition of a chalone the most important single characteristic is the tissue-specificity of the antimitotic action, although in practice this may be difficult to prove with complete certainty because of the large number of other tissues against which its action should ideally be tested.

However, if proof of the principle of chalone tissue-specificity is needed then the granulocyte and erythrocyte systems provide the perfect evidence. Both of these cell types are derived from the same cell population, both occupy the same spaces in the bone marrow and in the blood system, and both can be studied by the same assay methods. The results obtained from all the many experiments that have been carried out have always been the same: the granulocyte chalone inhibits mitosis only in the immature granulocytes, while the erythrocyte chalone inhibits mitosis only in the immature erythrocytes (Rytömaa, 1976a).

The chalone control of mitotic activity is normally exercised locally, and consequently, because the immature cells of these two tissues lie intermingled in the same narrow spaces, it is particularly important that their chalone control systems should be absolutely distinct. A similar situation exists in the skin, where such diverse tissues as the epidermis, sebaceous glands, sweat glands, melanocytes, etc. lie closely together and have sharply separate chalone control systems (see 10.2.1).

The Failure of Homeostasis

Eine neue, wissenschaftliche Wahrheit triumphiert nicht dadurch, dass sie ihre Gegner überzeugt und zur Einsicht bringt, sondern dadurch, dass ihre Gegner allmählich aussterben und eine neue Generation heranwächst, die mit ihr längst vertraut ist.

Max Planck, 1858–1947

Chapter 9

Chalones and cancer

9.1	Carcinogenesis	128
	9.1.1 The diversity of tumours	128
	9.1.2 The process of initiation	130
	9.1.3 The latent period and promotion	130
	9.1.4 The process of progression	132
	9.1.5 Undifferentiation and dedifferentiation	133
9.2	Chalones and cancer	134
	9.2.1 The chalone mechanism in tumours	134
	9.2.2 The tumour cell membrane	136
	9.2.3 The structure of tumours	136
	9.2.4 The sigmoid curve of tumour growth	138
	9.2.5 The plateau phenomenon	139
	9.2.6 Cell death in tumours	141
9.3	Chalones and tumour destruction	141
	9.3.1 The experimental evidence	142
	9.3.2 The clinical evidence	143
	9.3.3 The causes of tumour destruction	145
9.4	Summary and conclusions	147
	9.4.1 The nature of tumours	147
	9.4.2 Cancer research	148
	9.4.3 Historical notes	149

In all epithelial tissues cellular homeostasis is maintained in the same way through a chalone control mechanism which ensures complete stability over the whole range from extreme hypoplasia to extreme hyperplasia. In all connective tissues the same mechanism also operates, but with the added complication of constant cell recruitment from a multipotential stem cell population.

The question that now arises is whether a chalone control mechanism could break down, and if so what would be the consequences. In theory the mechanism could fail in either of two directions. In the first, cell loss might come to exceed cell gain so that the tissue would disappear. This happens when epidermis is separated from its dermis, and it may also happen naturally when, during early development, certain vestigial organs or structures are lost. It is also possible that, after a particular type of cell damage, it may develop pathologically in a small group of tissue cells, which would then simply die. This occurrence would, of course, not be noticed.

In the second place, cell gain might come to exceed cell loss, and if it continued it would be lethal. It happens naturally as a temporary phase during wound healing and compensatory hypertrophy; it develops pathologically, but also as a temporary phase, during the growth of a chronic tumour; and it persists pathologically and fatally during the growth of a lethal or malignant tumour.

It has been shown that the raised mitotic rate, and the consequent increase in cell number, seen during wound healing and in compensatory hypertrophy is due to the combination of a reduced chalone concentration and a reduced cell responsiveness to what chalone remains. The obvious question thus arises whether the same causes, in more extreme forms, may be responsible for tumour growth.

9.1 Carcinogenesis

Before attempting to answer this question, it is necessary to review briefly what is known of the origins of a tumour. In doing this it is important not to be confused by the sheer mass of interesting but essentially irrelevant information that has been accumulated.

9.1.1 *The diversity of tumours*

In particular the picture has been complicated and the situation obscured by the wide variety of tumour types that have been recognized and listed, as well as by the many and diverse causes that have been identified.

Regarding the variety of tumours, not only are different tumour types derived from different tissues, but as Foulds (1963) has emphasized, even tumours derived from the same tissue in the same animal by the same carcinogenic process may be 'extremely varied and integrated' so that 'it is probable that no two tumours are exactly alike in every respect'. Similarly, Berenblum (1967) has stated: 'Since every normal type of cell can give rise to a cancer cell, there must be at least as many different kinds of cancer as there are cell types in the body. This gives a measure of the complexity of the cancer problem . . . cancer is not one disease but a whole galaxy of separate diseases.'

The opposite is in fact the truth: whatever the tissue of origin and whatever the cause, the damage to the homeostatic control mechanism is always the same, leading simply to an excess of cell gain over cell loss. The histological appearance of a tumour is of itself unimportant although it may identify the tissue of origin, and by indicating the amount of ancillary damage that has been suffered, it may demonstrate how much of the normal tissue-specific characteristics of the cells have been modified or destroyed. However, such distortions of normal cell structure contribute nothing to the growth of the tumour.

Similarly, the nature of the diverse agents that are known to cause the specific type of cell damage needed to initiate tumour growth are, of themselves, unimportant. They fall into three main groups: physical agents, which include various forms of ionizing radiation; chemical agents, which include a wide range of seemingly unrelated substances; and biological agents, which include not only various viruses but also certain inherited cellular weaknesses. The recognition of all such factors may be important, especially in indicating which environmental factors should, if possible, be avoided, and an understanding of their modes of action will certainly be interesting. However, this is not of primary significance. It is the nature of the damage that has been inflicted on the cellular homeostatic mechanism that needs to be defined if attempts are to be made to counteract it, and not the manner in which the damage was inflicted.

Setting aside the apparent diversity of the cancer problem and concentrating instead on its evident unity, it is possible to recognize three main steps in the formation of any tumour. These are: initiation, whereby the appropriate damage is caused, and the tumour cells are created; promotion, whereby these cells, which may otherwise remain dormant, are stimulated to multiply into a visible tumour; and progression, whereby the tumour cells change their character, step by step, to become not only faster growing but also migratory within the body, and so ultimately fatal.

9.1.2 *The process of initiation*

Initiation is often said to be spontaneous, which merely means that the causative agent is unknown; in experimental conditions it is said to be induced, which means that the agent is known. The action of such a carcinogenic agent is not only qualitative but quantitative: with smaller doses fewer tumour cells are created and these, after promotion, are likely to produce only benign papillomata, which may regress; with larger doses more tumour cells are produced and these, after promotion, are likely to produce benign papillomata which progress into actively growing carcinomata; with still larger doses such carcinomata may be produced directly (Foulds, 1969).

The cell damage inflicted during initiation permanently changes the cell, which then passes its abnormal characteristics to its daughter cells. No reversion to the original normal state ever occurs. The change has thus the character of a mutation. It may be due to damage to some self-replicating part of the cell such as the outer membrane, when it is termed a somatic mutation, or to damage to the genes themselves, when it is called a genetic mutation, or it may be due to the distorting effects of additional genetic information provided by the DNA of invading viruses. It may also require not just one damaging episode but a particular pattern of accumulated damage, of which a part may even be inherited.

9.1.3 *The latent period and promotion*

Cancer cells, once formed, commonly remain dormant. This latent period may end quickly in the production of a visible tumour, or it may be so prolonged that the tumour never appears before the animal dies of old age. According to Foulds (1969), 'nearly all human beings who reach advanced old age harbour somewhere in their bodies at least one focus of latent carcinoma'. It has often been proposed that dormancy is maintained because the few damaged cells are isolated within the mass of normal cells, and so continue to be dominated by the growth control mechanism of these normal cells (Berenblum, 1954; Bullough, 1964). This would imply that the cancer cells are still capable of responding normally to the various messenger molecules that exist within the normal tissue (9.2.1).

At this time the cancer cells are also liable to be destroyed. First, if they are recognizably abnormal they may be attacked by the body defences, and especially by the immune system. Indeed it has been suggested that one main function of this system may be to eliminate tumour cells which, it has been postulated, may be constantly forming in the body. On this theory any tumour that develops does so because for some reason its cells do not evoke an adequate immune response. Second, if the cancer cells are derived from a tissue with a relatively high mitotic rate, such as the duodenal crypts or the epidermis, they may be carried away by the distal flow of post-mitotic cells, even though they themselves may not mature normally. Third, if the cells have been too severely damaged they may be unable to complete a mitotic division, and so may ultimately die.

Thus the development of a tumour from a group of dormant tumour cells is a matter of chance. The cells may survive or they may not; the survivors may emerge from their period of dormancy early, or late, or not at all.

In 1954 Berenblum suggested that a group of latent tumour cells can break its dormancy, and so begin to grow independently, as soon as its radius becomes so great that the innermost cells break free from the controlling influence of the surrounding normal cell mass. This implies the existence of a diffusion gradient into the cell group of some inhibitor such as a chalone (Bullough, 1964), which in epidermis is known to exert a discernible antimitotic action for not more than about 1 mm. The evidence suggests that in epidermis, in a tumour cell group with a radius approaching 0.5 mm, the central cells would begin to break free from the chalone control and to multiply.

The chance that this would happen would be increased by anything that favoured an increase in the latent cell mass; one such factor would be a reduced chalone efficiency leading to an increased mitotic rate. The chance that it would not happen would be increased by anything that had the opposite effect; one such factor would be an increased chalone efficiency leading to a reduced mitotic rate.

This suggestion can be tested experimentally in various ways, especially by inducing chronic changes in the mitotic rate. Thus, with the proviso that a raised mitotic rate may also cause the latent cells to be sloughed off, it is well known that it can shorten the latent period. This has been shown in epidermis by the mitotic stimulus of wounding or irritation; in liver during regeneration after partial hepatectomy; in mammary gland by the mitogenic hormones that cause hypertrophy and lactation; and even, in several types of tumours, by adrenalectomy which weakens chalone action. In all these cases the tumours appear sooner and more of them form (see Bullough, 1965, 1967).

However, it is the opposite effect that is the more interesting. The length of the latent period, in both spontaneous and induced tumours, is increased when the mitotic rate is chronically reduced. This effect may even be extended to the point that no tumours appear at all during the lifetime of the animal, although the continuing presence of tumour cells can be shown at any time by the application of a stimulating or promoting agent.

This is illustrated by two situations. First, primary tumours are relatively rare in such naturally low mitotic rate tissues as those of the kidney and liver, even though in the course of a lifetime mutations may build up to a high level in their cells (Curtis, 1963). Second, they may become equally rare in all tissues during a period of chronic hypoplasia. Such hypoplasia can be induced experimentally by chronic stress, which results in the excessive production of the stress hormones, or by the continued administration of hydrocortisone (see Bullough, 1965, 1967).

Experimental stress, delaying or preventing tumour appearance, has been induced by high or low temperatures, by excessive noise, and by other similar means, but by far the most extensive and significant results have been obtained by means of partial starvation. Underfed animals are so restless that their adrenal glands enlarge greatly, their rate of stress hormone secretion is greatly increased, their general mitotic rate is depressed (Bullough and Eisa, 1950), and the chances of tumour appearance are reduced almost to zero (Tannenbaum and Silverstone, 1957). Equally remarkable is the fact that with this treatment the animals remain more healthy and live for almost twice as long as do the normal fully-fed control animals.

These dramatic and well-established results point the way to one practical approach to cancer prevention, which even after some 40 years remains unexploited and indeed unconsidered.

9.1.4 *The process of progression*

Tumours, when they appear, show a wide range of different characteristics, but whatever these are they are not permanent. They tend to change progressively, slowly or quickly, in many small steps or in a few large ones, in such a way that the tumour mass grows more and more quickly while the tumour cells, progressively losing the visible characteristics of their parent tissue cells, tend to break away and spread throughout the body (metastasis). If the tumour was originally benign, this process of progression leads ultimately to rapid growth, to malignancy and to death.

The sudden steplike changes that characterize progression appear to be due to further mutations, somatic or genetic, as though the original induced mutation had predisposed the affected cells to a series of further spontaneous mutations. This reinforces the view that to produce a lethal tumour may require not one change in the cell control mechanism but a pattern of such changes. However, a tumour may often appear in which all the necessary changes are already complete and which *ab initio* is rapidly growing, malignant and lethal.

It is important also to emphasize that even a lethal tumour does not necessarily show a high rate of new cell production; it is lethal simply because the rate of cell gain continues to exceed the rate of cell loss until the animal dies. In such a tumour the mitotic rate may even be lower than that of the parent tissue, evidently because of damage to the general meta-bolic efficiency of the cells or more specifically to some part of the long and complex mitotic mechanism. Such damage could have been caused during initiation or progression but equally it could be due to attacks by the body's defence mechanisms, especially the immune system.

9.1.5 *Undifferentiation and dedifferentiation*

In view of statements in the literature, one question that must be con-sidered concerns the fundamental nature of the change that has occurred in a cancer cell. This is a question that is confused by a general lack of understanding of the term differentiation (1.1). Thus it is commonly held that tumour cells have become 'undifferentiated', which implies that they have ceased to belong to or be typical of their tissue of origin, or that they have become 'dedifferentiated', which implies that they have reverted towards an embryonic state. Both these beliefs are misconceived.

Regarding the first, the cells of many lethal tumours retain their ability to differentiate for mitosis and for tissue function with ageing in an essentially normal manner (Bullough and Deol, 1971b). Other tumours in which the cells, to a greater or lesser degree, have lost their ability for tissue function because of damage to their tissue operon, have obviously not lost their ability to differentiate for the other cell functions of mitosis and ageing. In any normal tissue a cell that is differentiated for mitosis remains as firmly part of that tissue as does any functional cell; this rule is not altered in a tumour.

Theoretically the converse situation is also possible, namely that the mitosis operon in tissue cells may be so damaged that only the state of tissue function remains a possibility. Such cells would ultimately age and die, but equally they would not be undifferentiated.

The second theory that tumour cells are dedifferentiated and that 'the true comparison should be with an embryonic cell' (Haddow, 1967) is clearly nonsense. The range of potentialities open to embryonic cells is obviously not regained in tumour cells, even though in odd cases some curiously atypical syntheses may occur.

The question of differentiation is irrelevant to that of carcinogenesis. Cancer cells are simply differentiated cells that have been damaged in such a way that they are more likely to remain in the mitotic cycle, and less likely to mature for some form of tissue function, whether normal or distorted.

9.2 Chalones and cancer

Thus, in essence, the problem of cancer can be reduced to the question as to which part, or parts, of the cellular homeostatic control mechanism are damaged or broken when the rate of cell gain begins to exceed the rate of cell loss. In the words of Roe (1969) a tumour develops because of 'a failure in the generation of the mitosis restraining mechanism, or in its transport to the cell, a failure in the cellular apparatus responsible for receiving the restraining mechanism, or a failure of the cell to act on the message even though it received it'. The evidence that this 'restraining mechanism' is in fact the tissue-specific chalone is now overwhelming, and this lends obvious urgency to studies of the condition of the chalone control mechanism in tumour cells.

9.2.1 *The chalone mechanism in tumours*

The first attempts to determine the state of the chalone mechanism in tumour cells were those of Bullough and Laurence (1968b), who studied a rapidly-growing transplantable epidermal carcinoma of the rabbit, and of Rytömaa and Kiviniemi (1968b), who worked with a similarly rapidly-growing transplantable granulocytic leukaemia of the rat. The results obtained were the same in both cases.

The tumour cells were found to contain less than 10% of the chalone content of normal tissue cells; much of the lost chalone was present in the blood, which, therefore, had a higher than normal chalone content even though the tumour cells were perhaps not synthesizing the normal amount of chalone; and, most significantly, when the chalone concentration was artificially raised, whether *in vitro* or *in vivo*, the tumour cells responded with a powerful mitotic inhibition. Furthermore, although the epidermal carcinoma had been transplanted from rabbit to rabbit for 30 years and was completely anaplastic, and although the leukaemia was equally abnormal, each tumour responded by mitotic inhibition only to the chalone of its tissue of origin. Tissue-specificity of response had not been lost.

The conclusions were that both the chalone producing and the chalone responding parts of the cellular homeostatic mechanism remained relatively normal, but that the cellular chalone concentration was too low. The suggestion made was that the basic fault may lie in the cell membrane which either allows too much chalone to escape from the cell, implying that the chalone receptors are intracellular, or which is so damaged that the chalone receptors on the outer cell surface are unable to hold adequate numbers of chalone molecules. Later work has tended to support this second suggestion.

Since these early studies a wide range of tumour types in several mammalian species, including man, have been investigated and have given essentially similar results. At the moment the list includes some 20 different tumours in five different species: epidermal carcinoma of rabbit (Bullough and Laurence, 1968b), mouse (Bullough and Deol, 1971a; Kariniemi and Rytömaa, 1976), and hamster (Elgjo and Hennings, 1971); cervix carcinoma of mouse (Okulov et al, 1977); melanoma of mouse and hamster (Bullough and Laurence, 1968c; Seiji et al, 1974); bronchiolar carcinoma of man (Houck, 1976; Korsgaard et al, 1978); colon carcinoma of man (Kanagalingam and Houck, 1976); mammary carcinoma of mouse, rat and man (Gonzalez and Verly, 1976; and see Iversen, 1981); plasmacytoma of mouse (Bichel, 1972); hepatoma of rat (Otsuka and Terayama, 1966); fibrosarcoma and osteosarcoma of man (Houck and Attallah, 1975); several types of lymphocytic leukaemia in mouse and man (Attallah and Houck, 1976; Houck and Patt, 1981) and of granulocytic (myeloid) leukaemia in rat and man (Rytömaa and Kiviniemi, 1968b, 1969, 1970; Rytömaa et al, 1976, 1977). The animal tumours were studied both in vitro and in vivo while the human tumours were all except one (9.3.2) studied in vitro. For a review of most of the present evidence see Iversen (1981).

These various investigations were carried out with varying degrees of thoroughness, but the important point is that none of them has contradicted the original conclusions. Of these the single most important fact that has been repeatedly confirmed is that all kinds of tumour cells seem able to respond tissue-specifically to the chalone of their tissue of origin when the concentration of this chalone is raised to a high enough level.

This conclusion adds weight to the belief that the factor preventing the multiplication of latent tumour cells is the chalone produced by the surrounding normal tissue cells, and that the action of a promoting agent in breaking the period of tumour cell dormancy is through a reduction in the chalone concentration or the chalone effectiveness.

9.2.2 *The tumour cell membrane*

The evidence has continued to support the view that the basic fault in a tumour cell may lie in the cell membrane. The structure of this self-replicating membrane may be damaged directly (somatic mutation), or the tissue-specific receptor molecules that are an essential part of the chalone mechanism may become inadequate or distorted (genetic mutation). The result may then be that too little chalone remains in or on the tumour cells.

There is a large literature promoting the theory that membrane damage is a significant feature of tumour cells (see Marx, 1974). Damage to the cell surface leading to a transient breakdown in mitotic control can also be achieved experimentally by the brief exposure of the cells to proteolytic enzymes which distort or destroy cell surface macromolecules, and it has been suggested that this may provide a model for what happens permanently in tumour cells. This also fits with a theory that a main protection against cancer may be provided by the immune system, the cells of which react to the surface abnormalities of the tumour cells. Also, as a tumour progresses from a benign to a lethal state, it is the surface abnormalities that ultimately enable the cells to break free from their tissue-specific anchorages and so to migrate around the body.

Thus Rubin (1966) has concluded that 'malignant transformation arises from a heritable disruption of the distribution and configuration of cell surface macromolecules', and Marx (1974) has reviewed evidence to suggest that cancer may develop when 'alterations in cell membrane composition affect the availability of receptors'.

Such an argument also links naturally with a comprehensive hypothesis, advanced for instance by de Terra (1974), that from the protozoans to the mammals, the cell surface in some way plays a critical role in the control of cell proliferation.

9.2.3 *The structure of tumours*

The growth of a solid tumour is basically an orderly process even though in its final form its structure may be chaotic. It certainly does not involve 'persistent, uncontrolled and haphazard cell multiplication' (Editorial, *J. Am. Med. Assoc.*, 1969); even the fastest-growing tumour develops according to a fixed pattern. Furthermore, a cancer is not capable of 'potentially unlimited growth' (Webster's Dictionary, 1963); its growth is self-inhibiting, and there is good reason to believe that even a lethal tumour would show only limited growth if its host could only have survived long enough (9.2.5).

The pattern of growth of a variety of rapidly growing, transplantable and malignant tumours was studied by Bullough and Deol (1971b), who agreed with Foulds' (1969) statement that 'tumours in general are not random conglomerations of cells but organized tissues with characteristic histological patterns'. The establishment of the basic pattern depends, first, on the connective-tissue-adjacent position of the mitotic cells, and second, on the distal movement, ageing and finally death of the post-mitotic cells. In any tumour these normal rules persist unchanged. When growth begins a solid tumour typically consists of an outer sheath of connective-tissue-adjacent mitotic cells, a medial sheath of post-mitotic ageing cells and an inner mass of dead cells. Except that the tumour forms a cyst instead of a sheet, the cells are stratified exactly as they are, for instance, in the epidermis. Furthermore, the pattern exactly resembles that seen in the early embryonic stage, for instance of the pancreas, in which all the mitotic activity is in the outermost connective-tissue-adjacent cells while all the inner cells are post-mitotic and maturing.

This standard pattern must in all cases have the same cause. In a tumour, as in a normal tissue, the antimitotic action of the G_1 chalone is weakest peripherally where the promitotic action of the mesenchymal factor is strongest; the G_1 chalone action is strongest in the central tumour region where post-mitotic ageing and death predominate.

This simple form of tumour structure is lost as the cell mass expands and becomes lobulated, and as the connective tissue with its capillary network ramifies between the lobules. Even at this time, however, it is often possible to see that the mitotic activity is connective-tissue-adjacent and this has led to another misconception. It is commonly held that mitotic activity in tumours is related not to the connective tissue but to the capillaries associated with that tissue, and thus to the food and oxygen provided by the capillaries. For the same reason it is also commonly believed that the non-mitotic state of the cells that are distant from the connective tissue is due simply to the fact that these cells are dying of starvation and hypoxia.

It cannot be too strongly emphasized that much, and in many cases most, of the cell death in tumours is due to the normal post-mitotic ageing process. However, when a tumour becomes large it is true that much of the central cell death may be related to vascular failure (Bullough and Deol, 1971b), although even here much normal post-mitotic cell ageing and death also occurs.

In general, the picture presented by a typical solid tumour is consistent with a situation in which the G1 chalone concentration is so low that there is an excessive number of mitotic-cycle cells and an inadequate number of post-mitotic ageing cells. Thus cell production continues to exceed cell loss.

9.2.4   The sigmoid curve of tumour growth

A growing tumour presents not only a pattern in space, a structural pattern, but also a pattern in time. It has been emphasized that the tumour cells continue to synthesize the chalone of their tissue of origin but that this chalone is lost at too high a rate into the circulation. This being so, the larger the tumour becomes the higher is the systemic chalone concentration, and this has two consequences.

The first is a steady reduction in the mitotic rate of the tumour's tissue of origin, and this has been repeatedly observed. Thus the growth of an epidermal carcinoma is accompanied by a progressive mitotic inhibition in the whole of the normal epidermis, but in no other tissue (Bullough and Deol, 1971a; Kariniemi and Rytömaa, 1976); an expanding granulocytic leukaemia depresses normal granulocyte production in the bone marrow, and also experimentally in an implanted diffusion chamber (Vilpo et al, 1973).

The second consequence is seen in the tumour itself. With continuing tumour growth, and with still higher systemic chalone concentrations, the tumour begins to inhibit itself, and this produces the well-known sigmoid curve of tumour growth. From one of the earliest analyses of tumour growth dynamics, Laird (1964, 1965) concluded that 'the growth of nearly all tumours reported in the literature is characterised by a continuous deceleration from the earliest period of observation', and that this 'retardation of growth appears to be an actively increasing depression of the specific growth rate, rather than a passive limitation imposed by the exhaustion of available growth-supporting factors in the environment'. Later Burns (1969) was even more specific in concluding that a tumour may regulate its own growth 'as normal and non-neoplastic tissues do by the production of a homologous specific mitotic inhibitor', and Bichel (1973) agreed that tumour growth, which he described as an exponential process limited by an exponential retardation, 'can be explained by the increasing concentration of a humoral inhibitor produced by the increasing number of tumor cells'.

By extensive experiments (9.2.5) Bichel showed that this inhibitor was specific to the type of tumour that produced it in that it inhibited mitotic activity only in that same type of tumour, and that it had the characteristics of a chalone.

The great significance of these conclusions, especially in the possibilities they offer for the inhibition of tumour growth clinically, have been almost entirely ignored. This is evidently because of an old and widespread belief that the observed slowing of tumour growth is explicable in terms of the 'major drain on the metabolic resources of the host', or of vascular failure as the newly-forming blood capillaries fail to keep up with the volume increase (Steel and Lamerton, 1966) leading especially to hypoxia (Hewitt et al, 1967), or to self-poisoning due to toxic breakdown products in the central necrotic mass, or even to an increasing immune response. Certainly some or all of these factors may help to inhibit the growth of many large tumours (Bullough and Deol, 1971b), but the evidence is clear that they are not the basic factors.

9.2.5   *The plateau phenomenon*

A tumour inhibits its own growth by its synthesis of a tissue- and tumour-specific mitotic inhibitor, which, as the tumour increases in mass, increases in concentration within the fixed limits of the body space. The consequence is that in a benign tumour growth ceases on a stable plateau when enough cells have been produced for self-inhibition, while in a lethal tumour the plateau is not reached before the animal dies.

The experimental evidence for this conclusion is extensive. When two identical tumours are present, the removal of one accelerates the growth of the other (Goodman, 1957); partial hepatectomy stimulates the growth of adenomatous hepatic nodules (Trotter, 1961); the amputation of a large primary tumour may cause the development of metastases that would otherwise not have appeared (Lewis and Cole, 1958; Schatten, 1958), which recalls the well-known clinical problem of metastatic growth after the removal of a primary tumour; in a large ascites tumour in which growth has stopped, active mitosis begins again when most of the free cells have been washed from the body cavity (Burns, 1969); and when two tumours of different tissue origins are present together in the same animal, they grow independently of each other so that each reaches its usual plateau as if the other was not present. This last experiment neatly disproves any idea that tumour growth is inhibited by nutrient exhaustion or by toxic metabolic products.

The most extensive studies of the plateau phenomenon in tumours are those of Bichel (1970, 1972, 1973), who carefully plotted the growth characteristics of three mouse ascites tumours, each derived from a different tissue and each reaching a stable plateau without killing its host. He found that the removal of tumour cells in the plateau phase causes the immediate resumption of growth of the remaining cell mass; that the cell-free ascites fluid, taken at the plateau phase and injected into another mouse, inhibits mitotic activity in tumour cells in the rapid growth phase, but only if these tumour cells are of the same type of tumour; that when two different tumours are grown simultaneously in the same mouse, each grows at its normal rate to reach its normal plateau irrespective of the presence of the other; and that, when two tumours are grown simultaneously in the same mouse, the cell-free ascites fluid from another mouse containing only one of the tumours inhibits the growth of only the same type of tumour leaving the other tumour to continue its uninhibited growth.

It was this last type of experiment that finally confirmed the important point that tumour growth is inhibited through the synthesis of a tumour-specific substance with the characteristics of a chalone. The fact that this substance is the chalone of the tissue from which the tumour was derived has been repeatedly shown by testing the actions of tumour extracts on the mitotic activity of their parent tissues (Bullough, 1975; Vilpo et al, 1973; and 9.2.4).

It is now obvious that the manner in which a benign tumour reaches a stable plateau closely parallels the manner in which, for example, a regenerating liver remnant reaches its stable plateau (6.2.3). In both cases active growth occurs because the systemic chalone concentration, and, therefore the cellular chalone concentration, is too low to prevent it, and in both cases growth ceases when the cell mass is great enough to raise the chalone concentration to an inhibiting level. The difference between the two cases is one of degree: while a normal systemic chalone concentration is all that is needed to stop normal regenerative growth, a higher-than-normal systemic chalone concentration is needed to stop tumour growth. A secondary difference is that while in normal regeneration growth will cease after a particular time (the time taken for the newly-formed post-mitotic cells to age and to begin to die) irrespective of the tissue mass, in a growing tumour this does not occur because the chalone concentration is so low that not enough of the cells can enter the post-mitotic phase.

9.2.6 *Cell death in tumours*

This emphasizes that the low chalone content of the tumour cells has two consequences: an increased cell tendency to remain in the mitotic cycle, and a decreased cell tendency to enter the post-mitotic ageing process. This second consequence is evidently the more important, and indeed it has often been proposed that tumour growth is not so much the consequence of an increased mitotic rate (which may not develop) as of a decreased potential of cell maturation. As already explained, the mitotic rate in a tumour may remain low, but if insufficient numbers of cells become post-mitotic to age and die, growth will continue inexorably. What happens when a tumour mass stabilizes at a plateau is partly a reduction in the mitotic rate, but more significantly an increase in the proportion of the cells that enter the ageing pathway, and so pass to their death. It is this that restores the balance between cell gain and cell loss.

Thus post-mitotic cell ageing continues apparently normally in tumours, showing that this part of the chalone control mechanism has also not been damaged. Furthermore, the normal link between the mitotic rate and the post-mitotic ageing rate seems to be unbroken. In fast-growing malignant tumours the extremely high mitotic rate is matched by an abnormally short post-mitotic cell life span; in an anaplastic epidermal carcinoma this life span may be reduced to only 2 or 3 days.

The other causes of cell loss from tumours have already been listed, and apart from the possibility of an inadequate blood supply (Bullough and Deol, 1971b), the most important of these are the lethal attacks of the immune system (Marx, 1974) and the killer cells (Marx, 1980). The role of these natural defences in clinical situations is emphasized in 9.3.3. The growth rate of any tumour then depends on the relationship between the rate of cell gain, however fast or slow this may be, and the rate of cell loss, which is the sum of the effects of all the contributing factors taken together.

9.3 Chalones and tumour destruction

Since no exception has yet been reported, the general conclusion must be that the tumour cells retain a relatively normal chalone control mechanism, the implication perhaps being that if this mechanism were to be significantly damaged the cells would not survive. All the various tumour cells so far studied continue to synthesize the normal chalone of their tissue of origin, and all continue to respond to it. Furthermore, that fundamental part of the chalone mechanism, the linkage between the mitotic rate and the rate of post-mitotic cell ageing, also remains functional, at least in relation to those cells that manage to become post-mitotic. Such cells may or may not continue to synthesize their normal tissue-specific cell products, but this is unimportant to cellular homeostasis and so has no influence one way or the other on tumour growth.

The only fault so far discovered in the chalone mechanism of tumours is the abnormally low chalone concentration in or around their cells. Since it is now well-established that an increase in the concentration of the relevant chalone inhibits the mitotic activity of tumour cells, it remains to consider the effect on tumour growth of continuing chalone treatment.

9.3.1 *The experimental evidence*

Experiments involving the repeated administration of extra chalone *in vivo* have not been numerous, because of the practical difficulty of obtaining large enough supplies of pure enough chalone extracts. The first experiments were those of Mohr *et al* (1968), who used partially purified pig skin extracts against two types of melanomata implanted in mice or hamsters, and of Rytömaa and Kiviniemi (1968b, 1970), who used a more highly purified granulocyte chalone preparation against a granuloma (a solid form granulocytic leukaemia) and a dispersed myeloid leukaemia implanted into young rats. The pig skin extracts contained a number of different tissue chalones including both the G_1 and G_2 melanocyte chalones; the other extracts contained only a granulocyte G_1 chalone (there being no granulocyte G_2 chalone), and no other tissue chalones were present. In all three sets of experiments the results were identical.

In the experiments on the mouse melanoma and the hamster melanoma, injections were given for only 5 days, but the results were dramatic. With optimum doses of skin extracts the tumours regressed and disappeared completely in the 75 treated mice (compared with no regressions in more than 1000 untreated control mice) and in 200 treated hamsters (compared with no regressions in more than 3000 untreated control hamsters). However, the short period of treatment was evidently inadequate, and in most of the mice and all of the hamsters the tumours later recurred. Nevertheless, 8 mice were completely cured and lived to an old age.

Although at the time it was considered unlikely that any factor other than the melanocyte chalone could have been responsible, it was later suggested that the real anti-tumour agent may have been the bacterium *Clostridium*, which may have contaminated the skin extracts (Mohr *et al*, 1972). There are several reasons why this is improbable, one being that careful tests made at the time showed no bacterial contamination, and another that the results obtained cannot be reproduced by using *Clostridium* alone (Bullough, 1975; Rytömaa, 1976b). Furthermore, in later experiments by Dewey (1977, personal communication), using the same mouse melanoma and similar, but strictly bacteria-free, skin extracts, the results were the same: the tumours regressed and disappeared.

More important are the more carefully planned and executed experiments of Rytömaa and Kiviniemi (1969, 1970), who used a rat chloro-leukaemia (granulocytic or myeloid leukaemia) in both its solid and dispersed forms. The granulocyte chalone extracts were highly purified and bacteria-free, and the injections, given subcutaneously, were continued for about 3 weeks.

With the solid granuloma the results obtained were exactly similar to those recorded for the two melanomata: the tumours became soft and necrotic, ulcerated through the skin, and in most cases disappeared. A high proportion of permanent cures was obtained, and allowing for a few spontaneous remissions in the control rats, the results were highly significant. Still more important, however, were the results obtained with the leukaemia in its dispersed form, since this type of tumour, when implanted into baby rats as was done here, never shows any spontaneous remissions. In all the chalone-treated rats the tumour growth was inhibited and in 10 out of the 40 rats used there was a permanent cure. The cured animals grew into adults, bred normally, and ultimately died of old age. With further studies to establish the optimum dosage and the optimum length of treatment it is probable that the percentage of cures could be increased.

9.3.2 *The clinical evidence*

The results obtained from these animal experiments have been criticized, and even dismissed on the grounds that a transplanted animal tumour does not behave like a spontaneous human tumour. Thus Iversen (1981) states that 'it is well known that the tumour-host relationship in transplanted tumours is very different from the tumour-host relationship in tumours primarily induced by viruses, irradiation or chemical carcinogens, or occurring spontaneously in a human being', and that 'tumour regressions or cures of transplanted experimental animal tumours are therefore not related to the cure of spontaneous or carcinogen-induced human tumours'. It follows that experiments on spontaneous tumours, if possible in man, are needed.

The difficulties in the way of human experiments are obvious, and the only information so far obtained relates once again to granulocytic (myeloid) leukaemia. For a long time clinical evidence has been available indicating the existence in man of a feedback mechanism that controls not only normal granulocyte production but also myeloid leukaemic cell production. As one example, it is known that the destruction of leukaemic cells by extra-corporeal irradiation produces a strong stimulus to further leukaemic cell proliferation (Chan and Hayhoe, 1971), and it now seems clear that this is a chalone effect.

The first evidence that granulocyte chalone inhibits the mitotic activity of human myeloid leukaemic cells came from the *in vitro* experiments of Rytömaa and Kiviniemi (1970). Since then this result has been repeatedly confirmed (Bøll *et al*, 1979), and in addition a similar result has been obtained *in vitro* using the lymphocyte chalone on human lymphocytic leukaemic cells (Houck and Attallah, 1975). Apart from ethical problems, the main difficulties in the way of repeating this kind of experiment *in vivo* have been, and still are, to obtain adequate supplies of active chalone extracts, and to purify these to a high enough degree to avoid the possibility of adverse side effects.

By 1975 it became possible by heroic efforts to produce enough granulocyte chalone in a pure enough state to allow Rytömaa *et al* (1976) to carry out the first clinical experiments, using volunteers with chronic or acute myeloid leukaemia. However, because the chalone supply was limited, only a short series of injections was possible in each case, and if only for this reason the magnitude of the response was unexpected. In the first patient the chalone dosage was estimated by extrapolation, weight for weight, from the successful rat experiments, but because of adverse side effects the treatment was discontinued after only 4 days. However, the leukaemic cells rapidly disappeared from both the blood and the bone marrow, and with no further treatment the patient lived a normal life for the next year.

Because of the side effects, which were later proved to be due not to the chalone but to certain impurities, the other 6 patients were given much lower chalone doses. These produced similar but lesser responses; one patient was estimated to have lost 3 kg of leukaemic cells in 2 weeks.

In all cases one critically important point was that the chalone injections suppressed only the granulocyte system and the leukaemic cells derived from it. All the other body tissues were unaffected so that, as the leukaemia regressed, there was a rapid and dramatic improvement in each patient's condition. This is in sharp contrast to the effects of conventional chemotherapy, which severely damages the whole body.

It has been stressed by Rytömaa *et al* (1977) that this was not a therapeutic trial, but merely a preliminary experiment designed to discover whether human spontaneous leukaemia will respond to the granulocyte chalone in the same way as does a rat transplanted leukaemia. The results show clearly that it will and that criticisms such as those of Iversen (1981) are invalid.

It is also remarkable that the average survival time of the 7 patients, all of whom received an inadequate chalone dosage, was about twice as long as would have been expected if they had received a conventional chemotherapeutic treatment. They were also spared the torture that such a treatment involves. Obviously this kind of study urgently needs to be repeated and expanded. On an experimental basis Rytömaa's results have already been confirmed by Bala and Kovalevskaia (1981) using patients with chronic and acute myeloid leukaemia, but what is now needed is a full-scale clinical trial.

Finally, one other potential use for chalones in the cancer field lies in their diagnostic value. In cases of leukaemia, the nature of which could not be determined, it has been possible to distinguish a myeloid leukaemia by its specific response *in vitro* to the granulocyte chalone, and a lymphocytic leukaemia by its specific response *in vitro* to the lymphocyte chalone. Similarly, Korsgaard *et al* (1978) have diagnosed an apparent epidermal carcinoma as a metastasis from another quite different tumour type; unlike the series of human epidermal carcinomata that were studied, it did not respond to the epidermal chalone.

9.3.3 *The causes of tumour destruction*

The complete cures, obtained with adequate chalone treatment in two types of cancer in animals, and the partial or full remissions, obtained with inadequate chalone treatment of human myeloid leukaemia, were unexpected. In theory it could have been anticipated that raising the systemic chalone concentration would, at best, cause a tumour to plateau at a smaller mass, with active growth being resumed when the chalone treatment ceased. Certainly a chalone is not toxic, and it could never be lethal to either normal or cancerous tissues. Some explanation of the results obtained is, therefore, needed.

It has been stressed that the rate of growth of any tumour is a function of two opposing factors: the rate of cell production, which is related to the degree of chalone lack, versus the rate of cell loss, caused partly by normal post-mitotic death (e.g. 17–34% of the tumour cells; Frankfurt, 1967), partly perhaps to a developing inadequacy of the blood supply, but most significantly to the attacks of the body's defence systems (Marx, 1974, 1980; Mizuno et al, 1975). When a tumour develops it is clear that the rate of cell gain by mitosis must exceed the rate of cell loss by all the various causes of cell death combined, and that the tumour will grow quickly or slowly according to the balance between these opposing factors.

It is this situation, in which the immune system and the killer cells are involved, that explains why a tumour can be destroyed, and not merely arrested, either when the rate of new cell production is sufficiently reduced or when the rate of cell death is sufficiently increased. An adequate chalone-induced reduction in the rate of tumour cell production must inevitably lead to the total disappearance of the tumour (Toivonen and Rytömaa, 1978; Rytömaa and Toivonen, 1979), partly because the stronger chalone action both limits the mitotic rate and increases the rate of post-mitotic cell death, but mainly because the body defences, which were previously too weak, are now able to prevail.

Such cellular destruction does not, of course, occur in the tumour's parent tissue because it is not subject to an immune attack. This tissue merely shrinks into hypoplasia from which it rapidly recovers when the chalone treatment ceases and the tumour mass has gone.

This is the present explanation of the results obtained with chalone treatment, and it is on this view that the usual chemotherapeutic techniques for cancer treatment are particularly to be condemned. Such treatment not only damages the tumour cells, but also destroys the cells of the defence systems which are of primary importance in combating tumour growth.

Indeed, it may well prove that an ideal form of cancer treatment may involve a combination of chalone therapy with an appropriate stimulus to the immune system. Inadvertently, such a stimulus may have been induced in the 7 patients with myeloid leukaemia in response to the residual impurities in the chalone preparation used.

146

9.4 Summary and conclusions

The nature of cancer and the methods by which it may best be treated are closely related. For too long cancer has been regarded as a medical problem to be met in the old ways with the knife, with radiation and with poison. Most recently the emphasis has been on poisons developed by biochemists who, working in the tradition of Paul Ehrlich, have synthesized stronger chemotherapeutic agents. Berenblum (1977) has properly queried whether 'such screening is justified at the present time – by trial and error without any theoretical basis or lead'.

In fact cancer has always been a biological problem requiring for its proper understanding a previous analysis of that homeostatic mechanism whereby, in any normal mitotic tissue, the two processes of cell gain and cell loss remain so perfectly balanced. The essence of the cancer problem lies not with its causes, which are many and various and which are mostly unavoidable in any population that survives into old age, but in the nature of the breakdown in cellular homeostasis, whereby cell gain comes to exceed cell loss. It is from an understanding of this that the most appropriate treatment may be devised.

9.4.1 *The nature of tumours*

The main conclusion reached here is that the cellular homeostatic mechanism in tumour cells is weakened but nevertheless remains operative. The mechanism whereby the mitotic process is initiated or inhibited is still functional, as also is the mechanism that initiates and controls post-mitotic cell ageing, and that maintains the close link between the mitotic rate and the ageing rate. The tumour cells still produce the chalone of their tissue of origin, although perhaps in reduced quantities, and they still respond to it, although perhaps in a lesser degree. Even when the tumour cells have lost all trace of their original tissue characteristics they continue to produce their tissue chalone, and to respond by mitotic inhibition only to that same chalone. In all the tumours so far studied it is simply the shortage of chalone in or on the cells that is the critical factor enabling them to continue to grow.

As it increases in mass a tumour produces increasing amounts of the chalone of its tissue of origin. The consequent rise in the systemic chalone concentration then forces the tissue of origin into hypoplasia and later may cause the tumour itself to stop growing. A lethal tumour, in which growth slows but does not cease, is simply one that would also have stopped growing if the host animal could have survived long enough.

In all this a tumour behaves like a regenerating liver, which also reaches a stable mass when the systemic concentration is high enough. Liver growth stops when the chalone concentration reaches its normal level; tumour growth stops when the chalone concentration reaches an abnormally high level. A second difference is that other factors, notably the immune system, also inhibit tumour growth. Thus the failure of a lethal tumour to stop growing is due partly to its failure to raise the systemic chalone concentration to a high enough level, and partly to its failure to provoke an adequate immune response. Also, after treatment by radiation or chemotherapy, the immune system itself may have been too severely damaged to function adequately.

9.4.2 *Cancer research*

Today there are three hopeful fields of biological research into the cancer problem that are related to the chalone concept. One of these fields has been explored but abandoned, one is active, and the third is only just beginning.

The abandoned field is that of cancer prevention, which was actively studied in the decade after 1940, especially by Tannenbaum and Silverstone (1957). Their biological technique for preventing the appearance of all forms of cancer, whether spontaneous or induced, was almost completely effective. This line of research died because those involved failed to understand the nature of the mechanism they had activated. With modern knowledge this study should be re-opened and re-appraised in terms of the effects of stress on the chalone mechanism. In particular, the roles of the two stress hormones should be analysed, and a search made for synthetic substitutes which would be equally effective but which would not cause unwanted or dangerous side effects.

The active field of research is that which attempts to exploit the belief that cancer cells may induce an immune response (Marx, 1974; Mizuno *et al*, 1975), and that when a tumour develops this response has not been adequate. 'The possible use of immunological methods in cancer is based on . . . experiments involving thymectomy and the injection of antilymphocytic serum, . . . on the evidence of specific antigens to autochthonous tumors in animals, and on the belief that immunotherapy might prove to be successful if the natural immunological reaction of the body against tumor cells could be artificially enhanced' (Berenblum, 1977). The possible combination of this approach with that of chalone therapy is obvious.

This leads to the third field of research based on the study of the chalone control mechanism. What is now urgently needed is the purification, the chemical analysis and finally the synthesis in quantity of some of the more important chalones. In the case of the granulocyte chalone this step is now almost complete (Balazs *et al*, 1980; Paukovits and Laerum, 1982; see 8.2.3). This will open the way, first, for a proper understanding of the functions of the G_1 and G_2 chalones and thus of the actual structure of the cellular homeostatic mechanism, and second, for the organization of full-scale clinical trials. In relation to cancer therapy these trials should include some consideration of the parallel role of the immune system; beyond cancer, they should explore the possible use of the lymphocyte chalone, or chalones, in the control of the rejection of transplanted tissues and organs.

If the chalones can be used therapeutically in these ways then they will possess one invaluable advantage over every other present day form of treatment, namely because they are non-toxic, tissue-specific, and tumour-specific in their actions they will leave all the other tissue systems of the body entirely unaffected. Their use will avoid the devastating side effects of the usual therapeutic agents.

9.4.3 *Historical notes*

Like most new ideas the chalone concept has its historical roots, especially in relation to cancer therapy. The idea that an extract of a normal tissue may contain something that will inhibit the growth of a tumour, derived from that same tissue, goes back for at least 50 years, and it might even be stretched much further back to Paracelsus with his dictum that 'similia similibus curantur'.

The use against cancer of crude tissue extracts, which probably contained what are now called chalones, began about 1930. In London, Baker (1933, 1935) was treating cancer patients with extracts of animal tissue 'corresponding to that of the primary growth in the patient', and after claiming encouraging results, he concluded that a 'carcinoma arises through the failure of a factor which normally inhibits the overgrowth of cells'. Simultaneously, in New York, Murphy (1935) and his co-workers at the Rockefeller Institute began a long series of animal studies using tissue extracts against experimental tumours. These finally led him to the conclusion that 'malignancy is due to the lack of or inadequacy of a normal factor'.

Interestingly, Murphy also reported that the idea of searching for a naturally occurring tumour inhibitor came from a *Times* editorial (18 July 1928), which was published in connection with a Cancer Congress then being held in London. The editorial noted that Murphy had reported at the Congress that certain substances extracted from tissues had the power to stimulate cancer growth, and it suggested that, this being so, it must follow that 'Nature possesses also growth-preventing substances'. The editorial added, somewhat critically, that research workers would be more likely to profit by searching for these latter.

However, by 1940 this line of seemingly promising research, which, as Murphy (1935) said, gave 'direction in the search for curative measures for cancer', stopped completely, evidently because neither the climate of opinion nor the available techniques were ready for it. The subject was not re-opened until 1957 when Osgood published his theoretical models of the leukaemias based on the failure of a tissue-specific negative feedback mechanism, and it was not developed until after the discovery of the chalones. It is to be hoped that the climate of opinion is at last ready to exploit the opportunity.

Envoi

Toutes choses sont dites déjà; mais comme personne n'écoute, il faut toujours recommencer.

André Gide, 1869–1951

Chapter 10

General summary

10.1	*Cellular homeostasis*	154
	10.1.1 *The mitosis control mechanism*	154
	10.1.2 *The ageing control mechanism*	156
	10.1.3 *The cell gain : cell loss linkage*	156
	10.1.4 *The structure of a chalone mechanism*	157
10.2	*Chalone tissue-specificity*	158
	10.2.1 *Specificity in space*	158
	10.2.2 *Specificity in time*	158
	10.2.3 *The stem cell chalone systems*	159
10.3	*Tissue structure and function*	160
	10.3.1 *The body tissues*	160
	10.3.2 *Differentiation and modulation*	161
	10.3.3 *Differentiation for death*	162
	10.3.4 *The breakdown of homeostasis*	163
10.4	*Envoi*	164

In any mammal or other complex metazoan the functional structure of the tissues is maintained by layer upon layer of homeostatic mechanisms, beginning at the chemical level and extending upwards to the cellular level, the tissue level, the organ level, the organism level, and even to the population level. Within each of these levels the homeostatic mechanism is partitioned into functional units, which respond to perturbations independently of the other parts of the mechanism, and so confer a considerable adaptive flexibility on the whole. Furthermore, the controls within each organizational level interact with those of other levels 'so that the higher level systems, so called, continue to include in their operation all the relevant activities of the lower-order ones' (Goodwin, 1976). The result is a highly complex and extremely versatile whole.

The control mechanisms at the cell and tissue levels considered here depend fundamentally on the responses of the genes to an array of specialized messenger molecules. As usual these mechanisms are partitioned into discrete functional compartments which can react individually to changing circumstances. Thus, although the processes of cell gain and cell loss are essentially the same in all tissues, they are controlled tissue-specifically through their chalone mechanisms.

Indeed without some such system of separate tissue control the tissues could not exist in their present form, and the evolution of the metazoans would have been severely restricted or even impossible. The chalone homeostatic mechanism is fundamental to the very existence of the mammalian type of tissue and organ. It is an interesting question whether a similar homeostatic mechanism exists in the higher plants (Evans *et al*, 1979).

10.1 Cellular homeostasis

The main points discussed in this book may now be briefly summarized. Cellular homeostasis in a mammalian tissue depends on the tissue-specific actions of chalone messenger molecules on non-tissue-specific cellular mechanisms. A chalone may be likened to a key that fits only one of a range of complex locks that are all of common design. The only other tissue-specific part of the control mechanism is that which initiates and maintains tissue function.

10.1.1 *The mitosis control mechanism*

The start of the mitotic process evidently depends on the triggering of an operator gene which then activates the mitosis operon, and so promotes the synthesis in proper sequence of the array of necessary enzymes. Once the operator gene has been activated, the whole chain response always passes to completion.

It is probable that in all tissues the messenger molecule that activates the operator gene is, like the gene itself, always the same, namely the mesenchymal factor synthesized by the connective tissue cells. In an epithelial tissue this is the only external factor to impinge on the chalone mechanism; all the other messenger molecules, whether of the primary or secondary messenger systems, are synthesized within the tissue cells themselves. In a connective tissue all the messenger molecules may be synthesized within the tissue cells.

In all tissues the resistance to the pro-mitotic action of the mesenchymal factor is provided by the anti-mitotic G_1 chalone. The mitotic rate is then a function of the relative strengths, or concentrations, of these two substances.

Such scant evidence as there is suggests that the receptor sites for the mesenchymal factor and the G_1 chalone, and perhaps also for the G_2 chalone, are on the cell surface, and that secondary messenger systems must therefore exist to carry the messages through the cytoplasm to the nucleus. Thus the tissue-specific part of a chalone mechanism may be mainly extra-cellular, while the non-tissue-specific part of the mechanism is mainly intra-cellular. That the intra-cellular part of the mechanism is non-tissue-specific is supported by evidence from fused cells. It is well known that when nuclei from two different tissue types, which are normally controlled by different chalone systems, are combined within the one cell, a stimulus to mitotis always results in both nuclei entering the mitotic process simultaneously. Similarly, the nuclei in a syncytium always divide at the same time.

It has been suggested that the receptor sites on the outer cell membrane are occupied competitively by the mesenchymal factor and the G_1 chalone, although it is not obvious how such tissue-specific sites could be equally attractive to a non-tissue-specific and a tissue-specific molecule. It is therefore possible that these molecules may have separate receptor sites, the response of the cell then depending on the relative numbers of each that are occupied, that is on the relative strengths of the messages reaching the nucleus.

In summary, the mitosis control compartment of the cellular homeostatic mechanism has the following main components: the non-tissue-specific mesenchymal factor from the connective tissue which activates the mitosis genes; the tissue-specific anti-mitotic G_1 chalone and its receptor sites; the non-tissue-specific secondary messenger systems whereby the messages are passed from the cell surface to the genes; and the non-tissue-specific sequential responses of the mitosis genes leading to the syntheses of the mitosis-promoting enzymes.

10.1.2 The ageing control mechanism

The nature of this second component part of the cellular homeostatic mechanism, which is intimately linked with cell maturation and tissue function, is obscure. In spite of its fundamental importance it has never been adequately studied. Three points are, however, already clear: first, that post-mitotic ageing is a dynamic and closely controlled non-tissue-specific process; second, that it is interlinked with the non-tissue-specific mitotic process; and third, that like the mitotic process it is tissue-specific-ally controlled.

The essential feature of the ageing process is progressive gene closure: first, there is the final closure of the mitosis operon; second, there is the final closure of all the remaining genes, which is sometimes combined with the death or disappearance of the nucleus; and finally, there is the short period of cell survival based on pre-formed messenger RNA and enzymes. This same sequence has been described in all cell types so far studied.

The rate of cell ageing is tissue-specific, and may change with changing conditions within the tissue. This indicates the existence of a tissue-specific control system that is evidently responsive to one or more types of messenger molecules. On inadequate grounds it has been suggested that the G_2 chalone may act as an inhibitor to control the ageing rate.

The main component parts of the ageing control mechanism can be only vaguely identified: the process is centred in the genes and is common to all tissues; the rate of cell ageing is influenced and apparently controlled by at least one tissue-specific messenger molecule; and the activation of the ageing process coincides with that of the tissue-specific genes on which tissue function depends.

10.1.3 The cell gain : cell loss linkage

An essential feature of a typical cellular homeostatic mechanism is the close connection that exists between mitosis control and post-mitotic ageing control. In any epithelial tissue, however widely the mitotic rate may change, the rate of post-mitotic cell ageing changes with it. The rate of cell gain is always exactly matched by the rate of cell loss, so that from hypoplasia to hyperplasia the tissue mass changes relatively little.

If the mitotic rate is controlled by the G_1 chalone and the post-mitotic ageing rate is controlled by the G_2 chalone this linkage could be easily explained. With a raised stress hormone concentration both cell gain and

cell loss would be equally inhibited, while after tissue damage, and consequent chalone loss, both cell gain and cell loss would be equally accelerated. Furthermore, these responses would be tissue-specific.

In the connective tissues it is not yet clear whether the mitotic rate and
the ageing rate are or are not interconnected. In blood tissues, which
seem to lack a G_2 chalone, the rate of cell gain is a composite of the rate of
new cell differentiation and the degree of mitotic activity in the newly-
recruited cells, while the rate of cell ageing may always remain relatively
constant. For the structural connective tissues no information is
available.

10.1.4 *The structure of a chalone mechanism*

A chalone control system comprises all the mechanisms described above.
In each tissue the activities of the cells are determined by the relative concentrations of the various messenger molecules in their immediate
environment. This leads to the consideration of the spatial distribution of
these molecules and thus of the spatial organization of the tissue cells.

In any epithelial tissue, however complex it may seem to be, the
organization is simple: the basal connective-tissue-adjacent cells, or in
some tissues the cells that are adjacent to a particular region of the
connective tissue, are inclined to mitotic activity, while the supra-basal
cells are involved in post-mitotic tissue function and ageing. This cellular
pattern is maintained by two opposing concentration gradients, the
mesenchymal factor concentration being highest basally, and the G_1
chalone concentration being highest supra-basally. The continuing
existence of the tissue depends on the continuing existence of these two
gradients.

This is a common pattern in dynamic biological systems. 'The simple
gradient hypothesis of the spatial organisation of a tissue is characteristic
of a whole class of developmental models for the explanation of pattern
formation and morphogenesis', and in all such models the messenger
molecules that control the 'changes in shape and form exert their
influence by some means similar to that proposed for chalone action'
(Goodwin, 1976).

However, in a connective tissue the situation is different because these
two concentration gradients are absent, the mesenchymal factor and the
G_1 chalone being evenly distributed throughout the whole tissue mass. A
typical chalone control mechanism cannot, therefore, exist and an
ancillary mechanism for controlled cell production is necessary. The

problem is solved by the constant recruitment of new cells by a terminal differentiation from a stem cell population, that is by the retention into adult life of a type of mechanism that is typical of embryonic and fetal life.

10.2 Chalone tissue-specificity

The tissue-specificity of the antimitotic actions of the chalones is a basic characteristic in the definition of this class of chemical messengers, and the question of the degree of specificity shown has been partly considered (8.3.4).

10.2.1 *Specificity in space*

The first generalization is that wherever different tissues are closely crowded together their chalone systems are sharply distinct. Thus mammalian skin consists of a wide range of different tissues, many of which are epidermal derivatives and so are closely related. These skin tissues either lie close together, as do the epidermis, dermis, hair follicles, sebaceous glands and sweat glands, or they are actually intermingled, as are the epidermal cells, the melanocytes and the Langerhans cells. All these tissues are either known to have strictly tissue-specific chalone systems (Iversen, 1981) or are suspected of having them.

A similar situation exists within the confines of the bone marrow, where the mitotic and mature cells of the various blood tissues lie closely packed together. All these tissues, as well as the stem cells from which they are derived, are either known or believed to synthesize strictly tissue-specific chalones (8.3.4).

However, where tissues are spaced widely apart a different situation may exist. Chalone control is normally exercised locally, and if tissues do not lie close together strict chalone tissue-specificity is not essential. One example is provided by the widely spaced epithelia of the epidermis, the mouth and tongue, the oesophagus, the trachea and the bronchioles, all of which seem to share the same G_1 and G_2 chalones, or alternatively to synthesize such similar chalones that each can act in a significant degree on all the other epithelia.

10.2.2 *Specificity in time*

Another question that arises is whether the chalones that exist in embryonic and fetal tissues are the same as those that operate in adult tissues. So far relatively little information exists on embryonic chalone systems, but enough is known to permit some preliminary conclusions.

In the first place there is good evidence that some embryonic tissues do respond to chalones extracted from their adult counterparts. One example is provided by an assay system that has been devised for the erythrocyte chalone, a method that depends on the inhibitory action of extracts of adult mammalian erythrocytes on the mitotic activity of the nucleated erythrocytes of 12–14 day chick embryos (Perrins and Jones, 1981). Another example is provided by the tissue-specific mitotic inhibition seen in the pronephric kidneys of amphibian tadpoles in response to treatment with extracts of adult rat metanephric kidneys (Chopra and Sinnett, 1971). Evidently in these systems little or no change occurs in the chalone systems from the embryo to the adult, and what is even more dramatic is that little or no change has occurred during the millions of years of vertebrate evolution that lie between the amphibians, the birds and the mammals.

However, in other tissues the situation is different. For instance, it has been noted that the epidermis of the newborn mouse does not respond to the G_1 chalone from adult epidermis. It only begins to respond after about 14 days, by which time all the many epidermis-derived tissues have been formed (2.3.4). Evidently the stratified epithelium that covers the newborn mouse is a proto-epidermis consisting of multipotential stem cells which differentiate not only into hair follicles and sweat glands but also into epidermis proper. When during development a tissue appears for the first time, one critically important step in its differentiation is the establishment of its tissue-specific chalone control mechanism. In mouse epidermis this occurs 14 days after birth.

Presumably the proto-epidermis of a mammalian fetus must possess its own specific chalone control system, and it would be interesting to know whether, during the annual antler regrowth in deer, the adult epidermis at the antler base reverts temporarily to the proto-epidermal condition as it recovers its ability to form new hair follicles.

10.2.3   *The stem cell chalone systems*

It is now clear that those stem cell populations that support the various connective tissues represent embryo-type mechanisms that have been retained into the adult stage to meet the special problem of connective tissue cell replacement. These stem cells are self-supporting in the manner of an epithelial tissue, gaining new cells by mitosis and losing mature cells by differentiation and possibly also by death. They may, therefore, be based on fully differentiated connective tissue to provide the mitotic stimulus, and the bone marrow stem cells are known to possess their own specific G_1 chalone (8.2.1); a G_2 chalone has not yet been sought.

The general conclusion is that in the embryo and fetus each multi-potential stem cell population must have its own chalone system. Some end-tissues, such as the kidney tubules, are formed early in embryonic life and are subject to adult-type chalone control from this time; others, such as the epidermis, are formed late in fetal life, or even after birth, and so acquire their adult-type chalone system much later; while still others, namely the various connective tissues, never lose their dependence on a stem cell population which, therefore, persists throughout life.

10.3 Tissue structure and function

The body of an adult mammal presents such a stable appearance that it would be easy to believe that, with all the complex developmental processes having been completed in the embryo and fetus, the tissues and organs require no further instructions in order to maintain their proper structure. However, it is now clear that the structure and function of adult tissues and organs must be constantly maintained by their genetic responses to gradients of specific inducing agents. The continuing existence of the tissues and organs must be constantly ensured by mechanisms of differentiation that are similar to, or identical with, those that are involved in organogenesis in the embryo. Of these mechanisms the chalone systems and the mesenchymal influences are of central importance.

10.3.1 *The body tissues*

From the present point of view there are four main types of body tissue: first, there are the epithelial tissues which are controlled by typical chalone mechanisms; second, there are the non-mitotic tissues which must have been subject to specific chalone mechanisms during their early development but in which new cell production has been reduced to zero and old cell death is minimal; third, there are the stem cell populations which have remained in an embryonic condition and which appear to be controlled by typical epithelial-type chalone mechanisms; and fourth, there are the connective tissues in which new cells are recruited partly by differentiation from stem cells and partly by the chalone-controlled multi-plication of the newly-differentiated cells.

160

10.3.2 *Differentiation and modulation*

All these tissue types, in the embryo, in the adult or in both, are in one way or another under some form of connective tissue control. This includes not only the usual stimulus to mitosis exerted by the mesenchymal factor but also a range of other important influences.

Mesenchymal inductions of tissues and organs, based on the activation of particular gene operons in response to specific messenger molecules, are a basic feature of embryonic development. Wessells (1977) has defined a set of rules that govern such differentiation. These include the following:

(1) The primary inducing substances are spatially arranged in the egg cytoplasm.

(2) The secondary inducing substances come into action following cell movements whereby the 'juxtaposition of previously separate cell populations allows the initiation of new tissue interactions'.

(3) The cell surface is the primary site for receiving and responding to developmental information, which may be either inhibiting or stimulating.

(4) Attachment to or contact with a solid base is an absolute requirement for many types of normal cell behaviour and development.

(5) These tissue interactions continue throughout post-embryonic life as sources of integration and stability for the continued maintenance of form and function.

The evidence reviewed here firmly supports this last rule. The embryonic types of cellular induction continue to operate throughout life, even though the responsive potentialities of the adult tissue cells are severely limited.

In addition to such genetic inductions the structural connective tissues continue to influence the state of the epithelial tissues by non-genetic means, and the resulting responses may be termed modulations. As one example, the regional differences in epidermal structure are created and maintained by such local factors as the strength of the dermal grip on the basal epidermal cells and the degree of folding of the dermal surface. In fact in all epithelia the area of the connective tissue surface determines the mass of the epithelium, and, therefore, of the organ of which it forms a part. The more connective-tissue-adjacent cells there are, the more supra-basal cells there are (the N ratio; see 3.1.1), and the more massive the tissue. In particular, an extensive three-dimensional connective tissue network, such as that in the liver, supports a particularly large epithelial cell mass.

However, the influence of the connective tissues on their adjacent epithelia may be even more complex than these various examples suggest. There may be other messages passing and other influences exerted over the epithelial–mesenchymal boundary that still remain to be identified. There are indications that a connective tissue may modify the nature of the epithelial cell's end-product, as for instance the precise form of the keratin synthesized by epidermis-derived cells, and that it may even influence the migration route followed by the post-mitotic epithelial cells, and thus the cellular architecture of the tissue and of the organ to which it belongs. Little or nothing is known of the nature of such influences, nor is much known of any reciprocal influences that may pass from an epithelium to its adjacent connective tissue. The whole question of epithelial-mesenchymal interactions remains one of considerable interest and importance.

10.3.3 *Differentiation for death*

Those tissue cells that differentiate for tissue function thereby also differentiate for death. Mature cell ageing and death is the necessary counterpart of new cell birth by mitosis, and mitosis is a process that must be retained in most tissues if only to make good the damage after any form of cell destruction or loss, or to allow for the development of any necessary form of tissue hyperplasia. The ageing and death of mature animals is also the necessary counterpart of the birth of a new generation, which not only replaces those individuals that have been destroyed but also allows for evolutionary experiment and progress.

These two ageing processes, at the cellular and at the body levels, show remarkable similarities. The post-mitotic life span of a mature cell is tissue-specifically determined, while the life span of any mammal is species-specifically determined. More important, it has been well established in mice and rats that the same physiological factor, a persistently high systemic concentration of the two stress hormones, greatly increases both the post-mitotic cellular life span and the animal life span.

The suggestion has been made that these two ageing mechanisms are interrelated (7.3.1), and specifically that the life span of any mammalian species is a function of the life span of the cells of the non-mitotic tissues. This may indeed be the reason why these tissues have become non-mitotic; a cellular life span mechanism may have become adapted to provide an animal life span mechanism. With no cell replacement possible the steady deterioration of the non-mitotic tissues must lead to the death of an animal following a time interval that is determined by the rate of this deterioration.

Whenever any physiological process requires for its success the co-operation of a number of different tissues and organs, it is common for only one tissue to act as initiator and pacemaker. In the case of animal ageing this would seem to be especially necessary and if a pacemaker tissue had to be consciously chosen then the nervous and muscular tissues would be obvious candidates. Any reduction in their efficiency would not only decrease the nervous responses and the muscular strength, but through the diminishing efficiency of the circulation it would come to involve all the body tissues. In the wild, death would occur when bodily efficiency fell below that needed to cope with the current circumstances; in domestic animals and in man the ageing process, carried much further, is ultimately seen in an exaggerated form, and is accompanied by patho-logical conditions, including cancer, that are unnatural in that they rarely occur in the wild.

10.3.4 *The breakdown of homeostasis*

In its extreme form the process of ageing leads to a wide variety of metabolic weaknesses and functional inadequacies, but from the point of view of cellular homeostasis, and perhaps in relation to a weakening of the body defence systems, the one important pathological development that is typical of advancing age is cancer. This potentially lethal form of cellular overgrowth also occurs, though much less frequently, at the other end of the life scale, when in young animals those cellular processes that are part of the pattern of tissue differentiation may become disorganized.

A typical tumour grows according to a well-defined pattern both in space and in time; in the words of Böhmig (1937), 'a tumour is no wild, unregulated, atypical growth but a well-organised differentiated whole that is to be distinguished only by its increased growth potential'. It grows in this controlled way because, although the chalone influence is weakened, the tumour cells remain in some degree responsive to all the messenger molecules that control the normal process of cellular homeo-stasis. A rise in the concentration of the chalone of the tumour's tissue of origin, whether induced artificially by injection or caused naturally by the tumour's own growth, results in an inhibition of mitosis in the tumour cells.

In other words, the activities of the tumour cells, like those of normal cells, are determined by the cellular environment. After the initial damage has been done the newly-formed tumour cells are often prevented by their environment from growing into a tumour (9.1.3); 'tumour development can be effectively interfered with long after the initial irreversible transformation from the normal cell to the dormant tumour cell has taken place, thus providing one of the first indications of the potentialities of tumour prevention' (Berenblum, 1964). This opportunity has never been exploited.

After a tumour has appeared it may be inhibited in its growth by self-induced changes in its environment, so that sometimes it may even disappear spontaneously. More important it has been clearly demonstrated by Rytömaa and his co-workers that chalone injections can cure at least one kind of cancer in rats (9.3.1), and that, with inadequate treatment, they can induce complete remission of the same kind of cancer in man (9.3.2). If the body defences are involved in combating and destroying the tumour cells, then it is only necessary to reduce the rate of new cell production far enough to ensure that the rate of cell destruction shall predominate. The disappearance of the cancer 'is an *automatic* outcome of a chalone-induced inhibition . . . it *must* happen if the current concept of chalone action is true' (Rytömaa and Toivonen, 1979), although of course cases may be found in which the body defences are inadequate, or in which they have been too severely damaged by conventional cancer therapy.

In a branch of science that is so lacking in new ideas it is almost unbelievable that the opportunities created by Rytömaa's careful, thorough, and logical work have not been energetically exploited. On present evidence, admittedly inadequate, the potential therapeutic uses of the chalones include: the suppression of any form of hyperplasia, including that associated with arteriosclerosis (7.1.1); the identification of the tissue of origin of a tumour when this is in doubt (9.3.2); the elimination of at least some forms of lethal tumours and the inhibition of others (9.3.1, 9.3.2); and, using the lymphocyte chalone, the inhibition of organ transplant rejection (8.2.7). A thorough investigation is now needed into the details of chalone control systems, of chalone chemistry and chalone synthesis, and of the possible practical uses of the chalones.

10.4   Envoi

In a previous analysis of the process of differentiation (Bullough, 1967) it was suggested that in all its various forms, including those in adult mammalian tissues, differentiation is always based on the same relatively simple mechanism, discovered in bacteria, by which specific operons are activated or inactivated in response to the presence or absence of specific messenger molecules in the cellular environment. More recently, it has become increasingly obvious, as would be expected, that the mechanisms of gene response in the eukaryotes, in which the DNA is complexed with histones to form chromosomes that are enclosed in a nuclear membrane, are considerably more complicated than are those of the prokaryotes, in which the DNA forms a naked loop. Marx (1981) has even suggested 'that regulation in nucleated cells may differ significantly from that in bacteria', but the evidence so far available suggests that, although it may be more complex, it involves no fundamentally new features. The one significant change, emphasized here, is that the messenger molecules in such animals as mammals are synthesized within the tissues for the specific roles that they play instead of, as in the bacteria, originating mostly in the external environment.

Just as the bacteria owe their continuing survival to the responses of their genes to these external messenger molecules, so also do the adult mammalian tissues owe their continuing existence to the responses of their genes to the various internal messenger molecules. Particularly important for tissue and organ survival are the counter gradients of the mesenchymal factor and the G_1 chalone, which operate either at first hand in the epithelial tissues or at second hand through a stem cell population in the connective tissues.

What is now needed is a considerable increase in knowledge of eukaryotic gene control (Maclean, 1976; Marx, 1981), of the nature and mode of action of such messenger molecules as the chalones (Balazs et al, 1980; Paukovits and Laerum, 1982), of epithelial–mesenchymal interrelations (Wessells, 1977), and of the control of intra-tissue cell migrations (Abercrombie, 1980; MacKenzie, 1980). Such increased knowledge would also strengthen the biological approach to the cancer problem, and would open the way for natural physiological forms of treatment to supersede the punishing therapeutic techniques of the present time.

One final question, on which no direct information is yet available, is whether the same type of tissue control mechanisms as those described here also exist in other metazoans. It is already clear that tissue-specific chalone mechanisms exist in all the vertebrate classes so that, for instance, the epidermal chalone of a fish is fully active against epidermal mitosis in man (Bullough et al, 1967). A comparative study of cellular homeostasis and dynamic tissue structure in the various metazoan phyla would be of particular interest.

References

Abercrombie, M. (1957). Localized formation of new tissue in an adult mammal. *Symp. Soc. Exp. Biol.*, **11**, 235

Abercrombie, M. (1980). The crawling movement of metazoan cells. *Proc. R. Soc. London*, **B207**, 129

Allen, T. D. and Potten, C. S. (1976). Significance of cell shape in tissue architecture. *Nature*, **264**, 545

Argyris, T. S. (1977). Kinetics of regression of epidermal hyperplasia in the skin of mice following abrasion. *Amer. J. Path.*, **88**, 575

Attallah, A. M. and Houck, J. C. (1976). Lymphocyte chalone. In Houck, J. C. (ed.) *Chalones*. (New York: Elsevier)

Baker, H. S. (1933). The treatment of cancer with connective tissue extract. *Lancet*, **2**, 643

Baker, H. S. (1935). Tissue extracts in the treatment of cancer. *Lancet*, **2**, 583

Bala, I. M. and Kovalevskaia, N. P. (1981). Role of chalones in therapy of leukemias. *Prob. Gematol. Pereliv. Krovi*, **26**, 12

Balazs, A. (1979). *Control of cell proliferation by endogenous inhibitors.* (Budapest: Academy of Sciences Publishing House)

Balazs, A., Sajgo, M., Klupp, T. and Kemeny, A. (1980). Purification of an endopeptide to homogeneity and the verification of its selective inhibitory action on myeloid cell proliferation. *Cell Biol. Int. Rep.*, **4**, 337

Bayliss, W. M. and Starling, E. H. (1902). The mechanism of pancreatic secretion. *J. Physiol.*, **28**, 325

Bellamy, D. (1968). Long-term action of prednisolone phosphate on a strain of short-lived mice. *Exp. Gerontol.*, **3**, 327

Berenblum, I. (1954). A speculative review. *Cancer Res.*, **14**, 471

Berenblum, I. (1964). The two-stage mechanism of carcinogenesis as an analytical tool. In Emmelot, P. and Mühlbock, O. (eds.) *Cellular control mechanisms and cancer.* (Amsterdam: Elsevier)

Berenblum, I. (1967). *Cancer research today.* (London: Pergamon)

Berenblum, I. (1977). Cancer research in historical perspective. *Cancer Res.*, **37**, 1

Bernstein, I. A., Sachs, L., Ball, R. and Walker, G. (1977). Regulation in epidermal differentiation. In *Biochemistry of cutaneous epidermal differentiation.* Seiji, M. and Bernstein, I. A. (eds.) (Tokyo: University Park Press)

Bertsch, S. and Marks, F. (1974). Lack of an effect of tumor-promoting phorbol esters and of epidermal G1 chalone on DNA synthesis in the epidermis of newborn mice. *Cancer Res.*, **34**, 3283

Bichel, P. (1970). Tumor growth inhibiting effect of JB-I ascitic fluid. *Eur. J. Cancer*, **6**, 291

Bichel, P. (1972). Specific growth regulation in three ascites tumours. *Eur. J. Cancer*, **8**, 167

Bichel, P. (1973). Self-limitation of ascites tumor growth. *Nat. Cancer Inst. Mon.*, **38**, 197

Billingham, R. E. and Silvers, W. K. (1968). Dermoepidermal interactions and epithelial specificity. In Fleischmajer, R. and Billingham, R. E. (eds.). *Epithelial-mesenchymal interactions*. (Baltimore: Williams and Wilkins)

Bishop, J., Favelukes, G., Schweet, R. and Russel, E. (1961). Control of specificity in haemoglobin synthesis. *Nature*, **191**, 1365

Böhmig, R. (1937). Die morphologisch fassbaren Wachstumsgesetze drüsenbildender Karzinome und ihre Metastasen. *Verh. Deutsch. Path. Ges.*, **30**, 329

Boll, I. T. M., Sterry, K. and Maurer, H. R. (1979). Evidence for a rat granulocyte chalone effect on the proliferation of normal human bone marrow and of myeloid leukemias. *Acta Haem.*, **61**, 130

Braun-Falco, O. and Christophers, E. (1974). Structural aspects of initial psoriatic lesions. *Arch. Derm. Forsch.*, **251**, 95

Briggaman, R. A. and Wheeler, C. E. (1968). Epidermal-dermal interactions in adult human skin: role of dermis in epidermal maintenance. *J. Invest. Derm.*, **51**, 454

Briggaman, R. A. and Wheeler, C. E. (1971). Epidermal-dermal interactions in adult skin. *J. Invest. Derm.*, **56**, 18

Brugal, G. (1976). Presence of intestinal chalones. In Cairnie, A. B., Lala, P. K. and Osmond, D. G. (eds.) *Stem cells of renewing cell populations*. (New York: Academic Press)

Bucher, N. L. R. (1963). Regeneration of mammalian liver. *Int. Rev. Cytol.*, **15**, 245

Bucher, N. L. R. and Swaffield, M. M. (1964). Rate of incorporation of labeled thymidine into DNA in regenerating rat liver. *Cancer Res.*, **24**, 1611

Bullough, W. S. (1946). Mitotic activity in the adult female mouse. *Philos. Trans. R. Soc.*, **B231**, 453

Bullough, W. S. (1948). Mitotic activity in the adult male mouse. *Proc. R. Soc. Lond.*, **B135**, 212

Bullough, W. S. (1949). Age and mitotic activity in the male mouse. *J. Exp. Biol.*, **26**, 261

Bullough, W. S. (1950). Mitotic activity in the tissues of dead mice. *Exp. Cell Res.*, **1**, 410

Bullough, W. S. (1952). The energy relations of mitotic activity. *Biol. Rev.*, **27**, 133

Bullough, W. S. (1962). The control of mitotic activity in adult mammalian tissues. *Biol. Rev.*, **37**, 307

Bullough, W. S. (1964). Growth regulation by tissue-specific factors. In Emmelot, P. and Mühlbock, O. (eds.) *Cellular control mechanisms and cancer*. (Amsterdam: Elsevier)

Bullough, W. S. (1965). Mitotic and functional homeostasis. *Cancer Res.*, **25**, 1683

Bullough, W. S. (1967). *The evolution of differentiation*. (London: Academic Press)

Bullough, W. S. (1969). Epithelial repair. In Dunphy, J. E. and van Winkle, W. (eds.) *Repair and regeneration*. (New York: McGraw-Hill)

Bullough, W. S. (1972). The control of epidermal thickness. *Brit. J. Derm.*, **87**, 187

Bullough, W. S. (1973). Ageing of mammals. *Zeit. Alternsforsch.*, **27**, 247

Bullough, W. S. (1975). Mitotic control in adult mammalian tissues. *Biol. Rev.*, **50**, 99

Bullough, W. S. and Deol, J. U. R. (1971a). Chalone-induced mitotic inhibition in the Hewitt keratinizing epidermal carcinoma of the mouse. *Eur. J. Cancer*, **7**, 425

REFERENCES

Bullough, W. S. and Deol, J. U. R. (1971b). The pattern of tumour growth. *Symp. Soc. Exp. Biol.*, **25**, 255

Bullough, W. S. and Deol, J. U. R. (1972). Chalone control of mitotic activity in the eccrine sweat glands. *Brit. J. Derm.*, **86**, 586

Bullough, W. S. and Deol, J. U. R. (1975). Dermo-epidermal adhesion and its effect on epidermal structure in the mouse. *Brit. J. Derm.*, **93**, 417

Bullough, W. S. and Ebling, F. J. (1952). Cell replacement in the epidermis and sebaceous glands of the mouse. *J. Anat.*, **86**, 29

Bullough, W. S. and Eisa, E. A. (1950). The effects of a graded series of restricted diets on epidermal mitotic activity in the mouse. *Brit. J. Cancer*, **4**, 321

Bullough, W. S. and Laurence, E. B. (1960a). The control of epidermal mitotic activity in the mouse. *Proc. R. Soc. Lond.*, **B151**, 517

Bullough, W. S. and Laurence, E. B. (1960b). The control of mitotic activity in mouse skin. Dermis and hypodermis. *Exp. Cell Res.* **21**, 394

Bullough, W. S. and Laurence, E. B. (1961). Stress and adrenaline in relation to the diurnal cycle of epidermal mitotic activity in adult male mice. *Proc. R. Soc. Lond.*, **B154**, 540

Bullough, W. S. and Laurence, E. B. (1964a). The production of epidermal cells. *Symp. Zoo. Soc. London*, **12**, 1

Bullough, W. S. and Laurence, E. B. (1964b). Mitotic control by internal secretion. *Exp. Cell Res.*, **33**, 176

Bullough, W. S. and Laurence, E. B. (1966). Accelerating and decelerating actions of adrenalin on epidermal mitotic activity. *Nature*, **210**, 715

Bullough, W. S. and Laurence, E. B. (1968a). The role of glucocorticoid hormones in the control of epidermal mitosis. *Cell Tiss. Kinet.*, **1**, 5

Bullough, W. S. and Laurence, E. B. (1968b). Control of mitosis in rabbit V x 2 epidermal tumours by means of the epidermal chalone. *Eur. J. Cancer*, **4**, 587

Bullough, W. S. and Laurence, E. B. (1968c). Control of mitosis in hamster melanomata by means of the melanocyte chalone. *Eur. J. Cancer*, **4**, 607

Bullough, W. S. and Laurence, E. B. (1970a). Chalone control of mitotic activity in sebaceous glands. *Cell Tiss. Kinet.*, **3**, 291

Bullough, W. S. and Laurence, E. B. (1970b). The lymphocyte chalone and its anti-mitotic action on a mouse lymphoma *in vitro*. *Eur. J. Cancer*, **6**, 525

Bullough, W. S. and Laurence, E. B. (1971). *Unpublished results*

Bullough, W. S. and Mitrani, E. (1976). An analysis of the epidermal chalone control mechanism. In Houck, J. C. (ed.) *Chalones*. (New York: Elsevier)

Bullough, W. S. and Mitrani, E. (1978). The significance of vertical mitosis in epidermis. *Brit. J. Derm.*, **99**, 603

Bullough, W. S. and van Oordt, G. J. (1950). The mitogenic actions of testosterone propionate and of oestrone on the epidermis. *Acta Endocrinol.*, **4**, 291

Bullough, W. S. and Stolze, J. U. R. (1978). A new form of cellular arrangement in guinea-pig ear epidermis. *Brit. J. Derm.*, **99**, 519

Bullough, W. S. and Stolze, J. U. R. (1981). *Unpublished results*.

Bullough, W. S., Hewett, C. L. and Laurence, E. B. (1964). The epidermal chalone: a preliminary attempt at isolation. *Exp. Cell Res.*, **36**, 192

Bullough, W. S., Laurence, E. B., Iversen, O. H. and Elgjo, K. (1967). The vertebrate epidermal chalone. *Nature*, **214**, 578

Burns, E. R. (1969). On the failure of self-inhibition of growth of tumours. *Growth*, **33**, 25

Caplan, A. I. and Ordahl, C. P. (1978). Irreversible gene repression model for control of development. *Science*, **201**, 120

169

Carney, D. H. and Cunningham, D. D. (1977). Initiation of chick cell division by trypsin action at the cell surface. *Nature*, **268**, 602

Chan, B. W. B. and Hayhoe, F. G. J. (1971). Changes in proliferative activity in marrow leukemic cells during and after extracorporeal irradiation of blood. *Blood*, **37**, 657

Chase, H. B. (1954). Growth of the hair. *Physiol. Rev.*, **34**, 113

Chopra, D. P. (1978). Proliferation and differentiation of human keratinocytes *in vitro*. *In Vitro*, **14**, 939

Chopra, D. P. and Simnett, J. D. (1971). Tissue-specific mitotic inhibition in the kidneys of embryonic grafts and partially nephrectomized host *Xenopus laevis*. *J. Embryol. Exp. Morph.*, **25**, 321

Christophers, E. (1971). Cellular architecture of the stratum corneum. *J. Invest Derm.*, **56**, 165

Christophers, E., Wolff, H. W. and Laurence, E. B. (1974). The formation of epidermal cell columns. *J. Invest. Derm.*, **62**, 555

Curtis, H. J. (1963). Biological mechanisms underlying the aging process. *Science*, **141**, 686

Desser-Wiest, L. (1974). Stimulation of DNA synthesis in rat liver by adrenalectomy. *J. Endocrin.*, **60**, 315

Dewey, D. L. (1977). *Personal communication*.

Durward, A. and Rudall, K. M. (1958). The vascularity and patterns of growth of hair follicles. In Montagna, W. and Ellis, R. A. (eds.) *The biology of hair growth*. (New York: Academic Press)

Ebling, F. J. (1957). The action of testosterone and oestradiol on the sebaceous glands and epidermis of the rat. *J. Embryol. Exp. Morph.*, **5**, 74

Ebling, F. J. (1963). Hormonal control of sebaceous glands in experimental animals. In Montagna, W., Ellis, R. A. and Silver, A. F. (eds.) *Advances in biology of skin*. Vol. 4. (New York: Pergamon Press)

Elgjo, K. (1975). Epidermal chalone and cyclic AMP. *J. Invest. Derm.*, **64**, 14

Elgjo, K. (1976). Epidermal chalone in experimental skin carcinogenesis. In Houck, J. C. (ed.) *Chalones*. (New York: Elsevier)

Elgjo, K. and Devik, F. (1978). Growth regulation in X-radiated mouse skin. *Int. J. Radiat. Biol.*, **34**, 119

Elgjo, K. and Hennings, H. (1971). Epidermal chalone and cell proliferation in a transplantable squamous cell carcinoma in hamsters. *Virch. Arch. B. Zellpath.*, **7**, 1

Elgjo, K., Laerum, O. D. and Edgehill, W. (1971). Growth regulation in mouse epidermis. I. G_2 inhibitor present in the basal cell layer. *Virch. Arch. B. Zellpath.*, **8**, 277

Elgjo, K., Laerum, O. D. and Edeghill, W. (1972). Growth regulation in mouse epidermis. II. G_1 inhibitor present in the differentiating cell layer. *Virch. Arch. B. Zellpath.*, **10**, 229

Endo, H., Sugimoto, M., Tajima, K., Kojima, A., Atsumi-Hirato, Y. and Obinata, A. (1977). Adrenocortical control of epidermal keratinization. In Seiji, H. and Bernstein, I. A (eds.) *Biochemistry of cutaneous epidermal differentiation*. (Tokyo: University Park Press)

Epifanova, O. I. and Terskikh, V. V. (1969). On the resting periods in the cell life cycle. *Cell Tiss. Kinet.*, **2**, 75

Etoh, H., Taguchi, Y. H. and Tabachnick, J. (1975). Movement of beta-irradiated epidermal basal cells to the spinous-granular layers in the absence of cell division. *J. Invest. Derm.*, **64**, 431

Evans, L. S., Almeida, M. S., Lynn, D. G. and Nakanishi, K. (1979). Chemical characteristics of a hormone that promotes arrest in G_2 in complex tissues. *Science*, **203**, 1122

Fitzgerald, M. J. T. (1977). New knowledge about the epidermis. *Ir. J. Med. Sci.*, **146**, 206

Florentin, R. A., Nam, S. C., Janakidevi, K., Lee, K. T., Reiner, J. M. and Thomas, W. A. (1973). Population dynamics of arterial smooth-muscle cells. *Arch. Path.*, **95**, 317

Foa, P. (1982). *Personal communication.*

Foulds, L. (1963). Some problems of differentiation and integration in neoplasia. In Harris, R. J. C. (ed.) *Biological organisation.* (New York: Academic Press)

Foulds, L. (1969). *Neoplastic development.* (London: Academic Press)

Frankfurt, O. S. (1967). Mitotic cycle and cell differentiation in squamous cell carcinomas. *Int. J. Cancer*, **2**, 304

Franks, L. M. (1963). The balance between function and proliferation in organ culture. *Nat. Cancer Inst. Mon.*, **2**, 83

Franks, L. M., Riddle, P. N., Carbonell, A. W. and Gey, G. O. (1970). Lack of growth capacity of adult human prostate epithelium mechanically separated from its stroma. *J. Path.*, **100**, 113

Franks, L. M., Wilson, P. D. and Whelan, R. D. (1974). Effects of age on total DNA and cell number in the mouse brain. *Gerontologia*, **20**, 21

Gehring, U. and Coffino, P. (1977). Independent mechanisms of cyclic AMP and glucocorticoid action. *Nature*, **268**, 167

Gelfant, S. and Candelas, G. C. (1972). Regulation of epidermal mitosis. *J. Invest. Derm.*, **59**, 7

Gibbs, H. F. (1941). A study of the post-natal development of the skin and hair of the mouse. *Anat. Rec.*, **80**, 61

Glücksmann, A. (1964). Cell turnover in the dermis. In Montagna, W. and Billingham, R. E. (eds.). *Advances in biology of skin.* Vol. 5. (New York: Pergamon)

Gonzalez, R. and Verly, W. G. (1976). Isolation of an inhibitor of DNA synthesis specific for normal and malignant mammary cells. *Proc. Nat. Acad. Sci.*, **73**, 2196

Goodman, G. J. (1957). Effects of one tumor upon the growth of another. *Proc. Amer. Assoc. Cancer Res.*, **2**, 207

Goodwin, B. C. (1976). *Analytical physiology of cells and developing organisms.* (London: Academic Press)

Gospodarowicz, D., Hirabayashi, K., Giguère, L. and Tauber, J-P. (1981). Factors controlling the proliferative rate of bovine vascular smooth muscle cells in culture. *J. Cell Biol.*, **89**, 568

Goss, R. J. (1963). Effects of maternal nephrectomy on foetal kidneys. *Nature*, **198**, 1108

Goss, R. J. (1972). Wound healing and antler regeneration. In Maibach, H. I. and Rovee, D. T. (eds.) *Epidermal wound healing.* (Chicago: Year Book Medical Publishers)

Green, H. (1977). Terminal differentiation of cultured human epidermal cells. *Cell*, **11**, 405

Green, H. (1978). Cyclic AMP in relation to proliferation of the epidermal cell. *Cell*, **15**, 801

Grüneberg, H. (1952). *The genetics of the mouse.* (The Hague: Martinus Nijhoff)

Gurdon, J. B. (1977). Egg cytoplasm and gene control in development. *Proc. R. Soc. London.*, **B198**, 211

Haddow, A. (1967). In Teir, H. and Rytömaa, T. (eds.) *Control of cellular growth in adult organisms.* (London: Academic Press)

Halprin, K. M. (1976). Cyclic nucleotides and epidermal cell proliferation. *J. Invest. Derm.*, **66**, 339

Hanks, C. T. (1978). Inhibition of connective tissue proliferation by dermal extract. *J. Invest. Derm.*, **71**, 172

Hardrup, T., Hulth, A. and Telhag, H. (1975). Scattered mitoses in mature joint cartilage in rabbits after local trauma. *Clin. Orthop.*, **113**, 246

Halpap, B. and Cremer, H. (1972). Autoradiographische Untersuchungen am Granulationsgewebe. *Virch. Arch. Zellpath.*, **10**, 145

Hennings, H., Elgjo, K. and Iversen, O. H. (1969). Delayed inhibition of epidermal DNA synthesis after injection of an aqueous skin extract (chalone). *Virch. Arch. Zellpath.*, **4**, 45

Hewitt, H. B., Chan, D. P.-S. and Blake, E. R. (1967). Survival curves for clonogenic cells of a murine keratinizing squamous carcinoma. *Int. J. Radiat. Biol.*, **12**, 535

Hodges, G. M. (1967). Stromal-epithelial interactions. In *Biology of the periodontium*. (London: Academic Press)

Holley, R. W. (1975). Control of growth of mammalian cells in cell culture. *Nature*, **258**, 487

Hondius Boldingh, W. and Laurence, E. B. (1968). Extraction, purification and preliminary characterisation of the epidermal chalone. *Eur. J. Biochem.*, **5**, 191

Houck, J. C. (1976). Putative bronchial chalone. In Houck, J. C. (ed.) *Chalones*. (New York: Elsevier)

Houck, J. C. (1978). Lymphocyte chalone. *J. Reticuloendothelial Soc.*, **24**, 571

Houck, J. C. and Attallah, A. M. (1975). Chalones and cancer. In Becker, F. F. (ed.) *Cancer*, Vol. 3. (New York: Plenum)

Houck, J. C. and Cheng, R. F. (1973). Isolation, purification and chemical characterization of the serum mitogen for diploid human fibroblasts. *J. Cell Physiol.*, **81**, 257

Houck, J. C. and Daugherty, W. F. (1974). *Chalones: a tissue-specific approach to mitotic control*. (New York: Medcom Press)

Houck, J. C. and Patt, L. M. (1981). Lymphocyte chalone. *Lymphokines*. Vol. 4. (New York: Academic Press)

Houck, J. C., Sharma, V. K. and Cheng, R. F. (1973). Fibroblast chalone and serum mitogen. *Nature*, **246**, 111

Humphreys, T., Penman, S. and Bell, E. (1964). The appearance of stable polysomes during the development of chick down feathers. *Biochem. Biophys. Res. Commun.*, **17**, 618

Igarashi, Y. and Yaoi, Y. (1975). Growth-enhancing protein obtained from cell surface of cultured fibroblasts. *Nature*, **245**, 248

Iizuka, H., Kamigaki, K., Nemoto, O., Aoyagi, T. and Miura, Y. (1980). Effects of hydrocortisone on the adrenaline-adenylate cyclase system of the skin. *Brit. J. Derm.*, **102**, 703

Isaksson-Forsen, G., Burton, D. R., Korsgaard, R., Elgjo, K. and Iversen, O. H. (1977). Partial purification of epidermal G2 chalone. *Virch. Arch. B. Zellpath.*, **26**, 97

Iversen, O. H. (1961). The regulation of cell numbers in epidermis. A cybernetic point of view. *Acta. Path. Microbiol. Scand.*, **148** (*Suppl.*), 91

Iversen, O. H. (1981). The chalones. In Baserga, R. (ed.) *Handbook of experimental pharmacology*. (Berlin: Springer-Verlag)

Iversen, O. H., Bhangoo, K. S. and Hansen, K. (1974). Control of epidermal cell renewal in the bat web. *Virch. Arch. B. Zellpath.*, **16**, 157

Jacob, F. and Monod, J. (1963). Genetic repression, allosteric inhibition, and cellular differentiation. In Locke, M. (ed.) *Cytodifferentiation and macromolecular synthesis*. (New York: Academic Press)

Jensen, E. V. (1962). On the mechanism of estrogen action. *Perspect. Biol. Med.*, **6**, 47

Johnson, H. A. and Erner, S. (1972). Neuron survival in the aging mouse. *Exp. Geront.*, **7**, 111

Kanagalingam, K. and Houck, J. C. (1976). Colon carcinoma: its genesis and chalone control. In Houck, J. C. (ed.) *Chalones*. (New York: Elsevier)

Karasek, M. A. and Charlton, M. E. (1971). Growth of postembryonic skin epithelial cells on collagen gels. *J. Invest. Derm.*, **56**, 205

Karasek, M. A. and Liu, S-C. (1977). Keratinisation of epidermal cells in cell culture. In Seigi, M. and Bernstein, I. A. (eds.) *Biochemistry of cutaneous epidermal differentiation.* (Tokyo: University Park Press)

Kariniemi, A.-L. and Rytömaa, T. (1976). Effect of the Hewitt keratinising epidermal carcinoma on cell proliferation in different organs. *Brit. J. Derm.*, **94**, 515

Kiistala, U. (1972). Dermo-epidermal separation. *Ann. Clin. Res.*, **4**, 236

Kivilaakso, E. and Rytömaa, T. (1971). Erythrocyte chalone. *Cell Tiss. Kinet.*, **4**, 1

Korsgaard, R., Iversen, O. H., Isaksson-Forsen, G. and Burton, D. R. (1978). Human epidermoid lung carcinomas *in vitro* react to epidermal G_2 chalone. *Cell Tiss. Kinet.*, **11**, 441

Laerum, O. D. and Maurer, H. R. (1973). Proliferation kinetics of myelopoietic cells and macrophages in diffusion chambers after treatment with granulocyte extracts. *Virch. Arch. B. Zellpath.*, **14**, 293

Laird, A. K. (1964). Dynamics of tumor growth. *Brit. J. Cancer*, **18**, 490

Laird, A. K. (1965). Dynamics of tumor growth. *Brit. J. Cancer*, **19**, 278

Lajtha, L. G. (1973). Review of leukocytes. *Natl. Cancer Inst. Mon.*, **38**, 111

Leun, G. C. van der, Lowe, L. B. and Beerens, E. G. J. (1974). The influence of skin temperature on dermal-epidermal adherence. *J. Invest. Derm.*, **62**, 42

Levine, M., Pictet, R. and Rutter, W. J. (1973). Control of cell proliferation and cyto-differentiation by a factor reacting with the cell surface. *Nature*, **246**, 49

Lewis, M. R. and Cole, W. H. (1958). Experimental increase of lung metastases after operative trauma. *Arch. Surg.*, **77**, 621

MacDonald, R. A. (1961). Lifespan of liver cells. *Arch. Int. Med.*, **107**, 335

MacKenzie, I. C. (1969). Ordered structure of the stratum corneum of mammalian skin. *Nature*, **222**, 881

MacKenzie, I. C. (1970). Relationship between mitosis and the ordered structure of the stratum corneum in mouse epidermis. *Nature*, **226**, 653

MacKenzie, I. C. (1972). The ordered structure of mammalian epidermis. In Maibach, H. I. and Rovee, D. T. (eds.) *Epidermal wound healing*. (Chicago; Year Book Medical Publishers)

MacKenzie, I. C. (1980). Spatial organisation and tissue architecture in normal epithelia. In MacKenzie, I. C., Dabelsteen, E. and Squier, C. A. (eds.) *Oral pre-malignancy*. (Iowa: University Press)

Maclean, N. (1976). *Control of gene expression*. (London: Academic Press)

McLoughlin, C. B. (1961). Importance of mesenchymal factors in differentiation of chick epidermis. *J. Embryol. Exp. Morph.*, **9**, 385

McMinn, R. H. H. (1969). *Tissue repair*. (London: Academic Press)

Marks, F. (1971). Direct evidence of two tissue-specific chalone-like factors regulating mitosis and DNA synthesis in mouse epidermis. *Hoppe-Seyler's Z. Physiol. Chem.*, **353**, 1273

Marks, F. (1973). A tissue-specific factor inhibiting DNA synthesis in mouse epidermis. *Nat. Cancer Inst. Mon.*, **38**, 79

Marks, F. (1975). Isolation of an endogenous inhibitor of epidermal DNA synthesis (G_1 chalone) from pig skin. *Hoppe-Seyler's Z. Physiol. Chem.*, **356**, 1989

Marks, F. (1976). The epidermal chalones. In Houck, J. C. (ed.) *Chalones*. (New York: Elsevier)

Marks, F. and Rebien, W. (1972). Cyclic AMP and theophyline inhibit epidermal mitosis in G2 phase. *Naturwiss.*, **59**, 41

Marks, F., Bertsch, S. and Schweizer, J. (1978). Homeostatic regulation of epidermal cell proliferation. *Bull. Cancer*, **65**, 207

Marx, J. L. (1974). Biochemistry of cancer cells: focus on the cell surface. *Science*, **183**, 1279

Marx, J. L. (1974). Tumour immunology. *Science*, **184**, 552 and 652

Marx, J. L. (1980). Natural killer cells help defend the body. *Science*, **210**, 624

Marx, J. L. (1981). Gene control puzzle begins to yield. *Science*, **212**, 653

Mathé, G. (1972). Lymphocyte inhibitors fulfilling the definition of chalones and immunosuppression. *Rev. Eur. Étud. Clin. Biol.*, **17**, 548

Maurer, H. R. (1975). Chalones. In Talwar, G. P. (ed.) *Regulation of growth and differentiated function in eukaryotic cells.* (New York: Raven Press)

Medawar, P. B. (1963). Definition of the immunologically competent cell. In Wolstenholme, G. E. W. and Knight, J. (eds.) *The immunologically competent cell.* (London: Churchill)

Menton, D. N. (1976). A minimum-surface mechanism to account for the organisation of cells into columns in the mammalian epidermis. *Amer. J. Anat.*, **145**, 1

Mercer, E. H. (1962). The cancer cell. *Brit. Med. Bull.*, **18**, 187

Mitrani, E. (1978). Possible role of connective tissue in epidermal neoplasia. *Brit. J. Derm.*, **99**, 233

Mizuno, D., Chihara, G., Fukuoka, F., Yamamoto, T. and Yamamura, Y. (1975). *Host defense against cancer and its potentiation.* (Tokyo: University of Tokyo Press)

Mohr, U., Althoff, J., Kinzel, V., Süss, R. and Volm, M. (1968). Melanoma regression induced by chalone. *Nature*, **220**, 138

Mohr, U., Hondius Boldingh, W. and Althoff, J. (1972). Identification of contaminating *Clostridium* spores as the oncolytic agent in some chalone preparations. *Cancer Res.*, **32**, 1117

Montagna, W. and van Scott, E. J. (1958). The anatomy of the hair follicle. In Montagna, W. and Ellis, R. A. (eds.) *The biology of hair growth.* (London: Academic Press).

Moorhead, J. F., Paraskova-Tchernozenska, E., Pirrie, A. J. and Hayes, C. (1969). Lymphoid inhibitor of human lymphocyte DNA synthesis and mitosis *in vitro. Nature*, **224**, 1207

Murphy, J. B. (1935). Experimental approach to the cancer problem. *Bull. Johns Hopkins Hosp.*, **56**, 1

Nome, O.(1975). Tissue specificity of the epidermal chalones. *Virch. Arch. B. Zellpath.*, **19**, 1

Okulov, V. B., Anisimov, V. N. and Azarova, M. A. (1977). Effect of epidermal chalones on growth of transplantable tumors. *Bull. Exp. Biol. Med..*, **84**, 1473

Okulov, V. B., Ketlinsky, E. A., Ratovitsky, E. A. and Kalinovsky, V. P. (1978). Purification and characteristics of epidermal G2 and G1 chalones. *Biochimia*, **43**, 971

Omar, A. and Krebs, A. (1975). Mode of dermal-epidermal adhesion. *Dermatologia*, **150**, 321

Osgood, E. E. (1957). A unifying concept of the etiology of the leukemias, lymphomas, and cancers. *J. Natl. Cancer Inst.*, **18**, 155

Otsuka, H. and Terayama, H. (1966). Inhibition of DNA synthesis in ascites hepatoma cells by normal liver extract. *Biochim. Biophys. Acta*, **123**, 274

Patt, L. M. and Houck, J. C. (1980). The incredible shrinking chalone. *FEBS Lett.*, **120**, 163

Paukovits, W. R. and Hinterberger, W. (1978). Molecular weight and some chemical properties of the granulocyte chalone. *Blut*, **37**, 7

Paukovits, W. R. and Laerum, O. D. (1982). Isolation and synthesis of a hemoregulatory peptide. *Z. Naturforsch.*, **37c**, 1297

Paukovits, W. R. and Paukovits, J. B. (1975) Mechanism of action of granulopoiesis inhibiting factor. *Exp. Path.*, **10**, 348

Peckham, B. and Kiekhofer, W. (1962) Cellular behavior in the vaginal epithelium of estrogen-treated rats. *Amer. J. Obstet. Gynecol.*, **83**, 1021

Perrins, D. J. D. and Jones, W. A. (1981) An assay for erythrocyte chalone using chick embryo red blood cells. *Acta Haem.*, **65**, 247

Potten, C. S. (1974). The epidermal proliferative unit. *Cell Tiss. Kinet.*, **7**, 77

Prescott, D. M. (1976). *Reproduction of eukaryotic cells.* (New York: Academic Press)

Pritchard, J. J. (1974). Growth and differentiation of bone and connective tissue. In Goldspink, G. (ed.) *Differentiation and growth of cells in vertebrate tissues.* (London: Chapman & Hall)

Ristow, H.-J. (1982). Stimulation of DNA synthesis in cultured mouse epidermal cells by human peripheral leukocytes. *J. Invest. Derm.*, **79**, 408

Roe, F. J. C. (1969). Mechanisms of carcinogenesis. In Bittar, E. E. and Bittar, N. (eds.) *The biological basis of medicine*, Vol. 5. (London: Academic Press)

Ronzio, R. A. and Rutter, W. J. (1973). Effects of a partially purified factor from chick embryos on macromolecular synthesis of embryonic pancreatic epithelia. *Devel. Biol.*, **30**, 307

Rubin, H. (1966). Fact and theory about the cell surface in carcinogenesis. In Locke, M. (ed.) *Major problems in developmental biology.* (London: Academic Press)

Rytömaa, T. (1976a). Biology of the granulocyte chalone. In Houck, J. C. (ed.) *Chalones.* (New York: Elsevier)

Rytömaa, T. (1976b). The chalone concept. *Int. Rev. Exp. Pathol.*, **16**, 155

Rytömaa, T. (1978). Chalones and blood cells. In Gordon, A. S., Silber, P. and LoBue, I. (eds.) *The year in haematology.* (New York: Plenum)

Rytömaa, T. and Kiviniemi, K. (1968a). The control of granulocyte production. *Cell Tiss. Kinet.*, **1**, 329 and 341

Rytömaa, T. and Kiviniemi, K. (1968b). Control of DNA duplication in rat chloro-leukaemia by means of the granulocyte chalone. *Eur. J. Cancer*, **4**, 595

Rytömaa, T. and Kiviniemi, K. (1969). Chloroma regression induced by the granulo-cytic chalone. *Nature*, **222**, 995

Rytömaa, T. and Kiviniemi, K. (1970). Regression of generalised leukaemia in rat induced by the granulocytic chalone. *Eur. J. Cancer*, **6**, 401

Rytömaa, T. and Toivonen, H. (1979). Chalones: concepts and results. *Mech. Ageing Dev.*, **9**, 471

Rytömaa, T., Vilpo, J. A., Levanto, A. and Jones, W. A. (1976). Effect of granulocyte chalone on acute and chronic granulocytic leukaemia in man. *Scand. J. Haem.* **27** *Suppl.*, 5

Rytömaa, T., Vilpo, J. A., Levanto, A. and Jones, W. A. (1977). Effect of granulocytic chalone on acute myeloid leukaemia in man. *Lancet*, **1**, 771

Saetren, H. (1956). A principle of auto-regulation of growth. *Exp. Cell Res.*, **11**, 229

Sassier, P. and Bergeron, M. (1978). Specific inhibition of cell proliferation in the mouse intestine. *Cell Tiss. Kinet.*, **11**, 641

Sauerborn, R., Balmain, A., Goerttler, K. and Stöhr, M. (1978). On the existence of 'arrested G2 cells' in mouse epidermis. *Cell Tiss. Kinet.*, **11**, 291

Schäfer, E. A. (1913). In Proceedings of the International Medical Congress. *Brit. Med. J.*, ii, 380

Schatten, W. E. (1958). An experimental study of post-operative metastases. *Cancer*, **11**, 455

Schmidt, W., Richter, J. and Geissler, R. (1974). Die dermo-epidermale Verbindung. *Z. Anat. Entwickl.*, **145**, 283

Scott, R. B. and Bell, E. (1964). Protein synthesis during development: control through messenger RNA. *Science*, **145**, 711

Scott, E. J. van and Ekel, T. M. (1963). Kinetics of hyperplasia in psoriasis. *Arch. Dermatol.*, **88**, 373

Seiji, M., Nakano, H., Akiba, H. and Kato, T. (1974). Inhibition of DNA synthesis in melanocytes by a melanoma extract. *J. Invest. Derm.*, **62**, 11

Shields, R. (1978). Further evidence for a random transition in the cell cycle. *Nature*, **273**, 755

Skerrow, C. J. (1978). Intercellular adhesion and its role in epidermal differentiation. *Invest. Cell Pathol.*, **1**, 23

Slavkin, H. C. (1974). Embryonic tooth formation. In Melcher, A. H. and Zarb, G. A. (eds.) *Oral Sciences Review*, Vol. 4. (Copenhagen: Munksgaard)

Slavkin, H. C. (1980). Epithelial–mesenchymal interactions in normal epithelial differentiation. In MacKenzie, I. C., Dabelsteen, E. and Squier, C. A. (eds.) *Oral premalignancy*. (Iowa: University of Iowa Press)

Smith, J. A. and Martin, L. (1973). Do cells cycle? *Proc. Natl. Acad. Sci.*, **70**, 1263

Squier, C. A. (1980). The stretching of mouse skin *in vivo*. *J. Invest. Derm.*, **74**, 68

Starling, E. H. (1905). The chemical correlation of the functions of the body. *Lancet*, **ii**, 339

Steel, G. G. and Lamerton, L. F. (1966). The growth of human tumours. *Brit. J. Cancer*, **20**, 74

Stewart, F. A., Denekamp, J. and Hirst, D. G. (1980). Proliferation kinetics of the mouse bladder after irradiation. *Cell Tiss. Kinet.*, **13**, 75

Sun, T.-T. and Green, H. (1977). Cultured epithelial cells of cornea, conjunctiva and skin. *Nature*, **269**, 489

Tannenbaum, A. and Silverstone, H. (1957). Nutrition and the genesis of tumors. In Raven, R. W. (ed.). *Cancer*, Vol. 1. (London: Butterworth)

Terra, N. de (1974). Cortical control of cell division. *Science*, **184**, 530

Thomas, W. A., Janakidevi, K., Florentin, R. A., Lee, K. T. and Reiner, J. M. (1976). Search for arterial smooth muscle cell chalone. In Houck, J. C. (ed.) *Chalones*.(New York: Elsevier)

Toivonen, H. and Rytömaa, T. (1978). Monte Carlo simulation of malignant growth. *J. Theor. Biol.*, **72**, 257

Tosti, A., Nazzaro, P. and Porro, M. (1969). Histodynamics of the neoplastic and pseudoneoplastic growth of epidermis. *Ital. Gen. Rev. Derm.*, **9**, 1

Tosti, A., Scerrato, R. and Fazzini, M. L. (1959). Saggio di esplorazione biometrica dell'epidermide umano. *Ann. Ital. Der. Sif.*, **14**, 185

Trotter, N. L. (1961). The effect of partial hepatectomy on subcutaneously transplanted hepatomas in mice. *Cancer Res.*, **21**, 778

Vaughan, F. L. and Bernstein, I. A. (1971). Studies of proliferative capabilities in isolated epidermal basal and differentiated cells. *J. Invest. Derm.*, **56**, 454

Videman, T., Vilpo, L. and Vilpo, J. A. (1978). Inhibition of osteogenesis by bone mass in diffusion chamber cultures. *Exp. Path.*, **15**, 182

Vilpo, J. A. (1979). Non-specific and specific effects of a granulocytic bio-inhibitor. *Scand. J. Haem.*, **22**, 433

Vilpo, J. A., Kiviniemi, K. and Rytömaa, T. (1973). Inhibition of granulopoiesis by endogenous granulocyte chalone. *Eur. J. Cancer*, **9**, 515

Weiss, P. and Kavanau, J. L. (1957). A model of growth and growth control in mathematical terms. *J. Gen. Physiol.*, **41**, 1

Wessells, N. K. (1962). Tissue interactions during skin histodifferentiation. *Dev. Biol.*, **4**, 87

Wessells, N. K. (1964a). DNA synthesis, mitosis, and differentiation in pancreatic acinar cells *in vitro*. *J. Cell Biol.*, **20,** 415

Wessells, N. K. (1964b). Tissue interactions and cytodifferentiation. *J. Exp. Zool.*, **157,** 139

Wessells, N. K. (1977). *Tissue interactions and development.* (California: W. A. Benjamin)

Winter, G. D. (1972). Epidermal regeneration studied in the domestic pig. In Maibach, H. I. and Rovee, D. T. (eds.) *Epidermal wound healing.* (Chicago: Year Book Medical Publ.)

Winter, G. D. and Simpson, B. J. (1969). Heterotopic bone formed in a synthetic sponge in the skin of young pigs. *Nature*, **223,** 88

Index

Adrenalin: mitotic inhibition   32
Ageing: of animals   104, 162
    of cells   7, 34, 53
    of tumour cells   137, 141
Ageing pathway: control of   34, 104, 156
    phases of   21, 89
Androgens: mitotic stimulation   93
Antigens: induction by   118, 122

Blood tissues   111
    origins of   112
Body defences: against cancer   145

Cancer   127
    cell membrane and   136
    chalones and   134, 148, 163
    chalone diagnosis   145
    destruction   134, 141, 164
    diversity   128
    dormancy   130
    immune response   146, 148
    initiation   130
    nature of   147
    pattern of growth   136, 138, 163
    prevention   132, 148, 164
    progression   132
    promotion   131
    stress and   132, 148
Carcinogenesis   128
Cell death   7, 34, 53, 162
    in tumours   141, 146
Cell gain : cell loss ratio   41, 52, 97, 156

Cell migration   10, 72
Cellular homeostasis   7, 88, 121, 154
    breakdown of   11, 128, 147, 163
Cellular organisation   10, 54, 95
    basal cells   57
    supra-basal cells   63
    wounding and   67
Chalones   9
    cancer and   12, 127, 134, 163
    control mechanism   35, 52, 99, 154, 157
    chemistry   87, 115, 119
    definition   10
    in embryos   98, 159
    history of   8, 9, 149
    negative feedback mechanism   17, 50, 91, 147
    potential uses   164
    systems of   86, 157, 159
    tissue specificity   123, 158
    tumour destruction   141
Compensatory hypertrophy   90, 140
Connective tissues   106
    in blood   111
    cellular homeostasis   121
    chalone systems   108, 115, 116, 118, 121
    dermal cell types   107
    and differentiation   75, 121, 161
    stem cells   110, 112, 121, 159
    structural tissues   107

Death: in animals   104, 162
    in cells   7, 34, 53, 162

Defences against cancer 146
Dermis: cell types 107
  and epidermal induction 75, 161
  and epidermal modulation 78
  in hair root 85
  mesenchymal factor 23
  mitotic activity 108, 109
  stem cells 110
  wound healing 109
Dermo-epidermal junction 60
  relations 74, 79
Diet: and cancer 132
Differentiation 4
  in adult tissues 6, 76
  in bacteria 4
  basic rules 161
  connective-tissue-induced 161
  for death 162
  definition of 5
  dermis-induced 75, 79
  in embryo and fetus 75, 76
Duodenal mucosa: cell movement 96
  mitotic activity 85

Epidermal cell competence 38
Epidermis: basal cell extrusion 57
  basal cell grip 56, 58
  cellular homeostasis 35, 40
  cellular organisation 54, 70
  G1 chalone 25
  G2 chalone 31
  hypoplasia phase 1 42
  hyperplasia phase 2 44
  neonatal 29
  structure 55
  thickness 55, 70
  wound healing 45, 67
Epithelial tissues 83
Erythrocyte system 113, 116
  G1 chalone 116

Glucocorticoid hormone 27
  inhibiting mitosis 27
  promoting maturation 28
Granulocyte system 114
  G1 chalone 115

Haemopoietic system 111
Hair growth 85, 93, 95
  on antlers 6, 77
Homeostasis 7, 88, 121, 154
  levels of control 153

Hyperplasia 44
  and tension 59
Hypertrophy: compensatory 90
  hormone-induced 93
  periodic 92
Hypoplasia 42

Immune system: and cancer 146
  chalone system 118

Kidney: compensatory hypertrophy 90

Leukaemia: in rats 142
  in man 144
Liver: cellular organisation 96
  compensatory hypertrophy 90
  mass 99
Lymphocyte system: 118
  and cancer 146
  chalones of 119
  and tissue transplantation 120

Macrophage system 117
Maturation: of epidermal cells 28
Megakaryocyte system 118
Melanoma: destruction of 142
Mesenchymal factor 23, 84, 154
  chemistry 24
Mitosis: control 8, 23, 25, 36, 86, 121,
    154
  misconceptions on control 12
  operon 6, 154
  phases of cycle 19
  probability theory 38
  vertical axis 61
Muscles: smooth 101
  striped and cardiac 101
Myeloid leukaemia: destruction of 143,
    144

Negative feedback mechanism
  chalone 50
  criticism 18
  and tumour inhibition 138, 139
Neurones 103
Non-mitotic tissues 100

Oestrogens: mitotic stimulation 92

Plasma cells 118
Plateau phenomenon 90
  in tumours 139

Poietins   112, 122

Ratio constants   41
  number ratio   42, 70, 99
  rate ratio   41, 52, 70, 98

Sebaceous glands   95
Stem cells   120, 159
  of blood tissues   112
  of connective tissues   107, 110
  of lymphocyte system   118
Stratum corneum   67, 72
Stress and ageing   49, 104, 162
  and mitosis   48
  inhibition of tumours   132, 148
Structural connective tissues   107
Sweat glands   95

Tension: in epidermis   58, 62, 70
  in hyperplasia   59, 71

Tissue: mass   98, 161
  structure   55, 98, 160
  transplant rejection   120, 164
  types of   160
Tumour: cell death   137, 141, 146
  cell membrane damage   136
  destruction of   142, 143, 145
  diversity of   128
  growth of   136, 138
  initiation   130
  latent period   130
  plateau phenomenon   139
  progression   132
  promotion   131
  self-inhibition   138, 139
  structure   136

Vagina: hypertrophy   93

Wound healing: 45, 67, 90, 109
  hormone   8